THE AUSTRALIAN DEFENCE FORCE

MEETING THE MODERNIZATION CHALLENGES

ROBBIN LAIRD

SLDinfo.com

Library of Congress Control Number: 2024920839

The cover photo is credited to Dreamtime.com

ID 2280473 | Australian Map © Yong Hian Lim | Dreamstime.com

This book is dedicated to Catherine Scott of The Sir Richard Williams Foundation for her support for my work with The Sir Richard Williams Foundation

FOREWORD

The indefatigable Robbin Laird is at it again, providing a unique service to Australian defence and to the U.S.-Australia alliance, with enduring insights arising from the 2024 Williams Foundation seminar and beyond.

In *The Australian Defence Force: Meeting the Modernization Challenges*, Robbin Laird updates us on the work covered in his previous book, *Australia and Indo-Pacific Defence: Anchoring a Way Ahead* (2023). That work provided much more than just an overview of material covered at the previous Williams Foundation seminar. His lucid distillation of the key issues and challenges related to Australia and Indo-Pacific defence, spanning air, land, sea and beyond, warrant re-reading. It also sets up the reader for the contrast which follows in this volume.

Laird's new work focuses on the rub points arising from trying to implement long term strategic choices for acquiring next generation military capabilities, while being required to manage ongoing day-to-day operational requirements for the Australian Defence Force – a force which finds itself increasingly challenged. Laird explains this as 'a case study in the clash between force design for an envisaged force and the need to enhance the force in being to deal with the world as it is.'

This book highlights the enduring challenge faced by a democracy in getting the politics right while facing short political cycles and competing national priorities which inhibit government willingness to spend on defence. This reflects the age old 'guns' versus 'butter' dilemma faced by governments elected not on what might happen on the international stage, but on what they have done so far on the domestic front.

Laird captures the essence of the challenge for defence readiness whereby exquisite platforms are no longer the sole focus, and in planning for continent-spanning defence capabilities, how to build combat mass effectively. The challenge is a significant one in light of the scale of the emergent security issues in the Indo-Pacific to which Australia may be required to respond and the reluctance of governments to spend to match their rhetoric with reality.

Conflicts in the Middle East and Ukraine, and elsewhere, illustrate how, as he puts it, 'we have entered the world of the kill web where the evolution of warfare is being shaped by payloads' and 'combat clusters rather than platform identified task forces'. Uncrewed air, land and maritime vessels are transforming the way battles are fought. Along the way, the introduction of one new technological measure leads to the rapid development, prototyping and fielding of a countermeasure that, in turn, generates rapidly introduced counter-countermeasures. As a result, Laird explains, 'The problem is that for the current force to be effective, lethal and survivable, it needs to upgrade its force on the fly.'

The 2024 Williams Foundation seminar, which this book draws on, includes a wide range of perspectives on defence modernization and the need to adapt to the rapidly changing operational environment. These includes presentations from senior practitioners and service chiefs from the Australian Navy, Army and Air Force, as well as the space and cyber domains, along with Australia's AUKUS partner nations, the United States (General Gavin Schneider,

Commander U.S. Pacific Air Forces) and United Kingdom (Air Vice Marshal Mark Flewin).

What this amounts to is not so much a call for another revolution in military affairs, but a call for accelerated evolution: 'We don't need to destroy everything we've learned in the past. We just need to keep investing in that next stage'.

The Williams foundation drew in contributions from a stellar line-up of leaders, thinkers and writers on Australian defence and security, including Vice Admiral Tim Barrett, Air Marshal Stephen Chappell (Chief of Air Force) , Air Vice Marshal Glen Braz (current Air Commander, Australia), and Lieutenant Generals Susan Coyle (Chief of Joint Capabilities) and Simon Stuart (Chief of Army), as well as former director of the Australian Strategic Policy Institute, Peter Jennings, and Phil Winzenberg, the Deputy-Director General Signals Intelligence and Effects from the Australian Signals Directorate.

Winzenberg noted that 'The Five Eyes alliance is the greatest intelligence partnership the world has ever known... The trust and depth of the partnership... of what we do together is truly staggering.'

This volume also includes a number of significant interviews conducted by Laird covering a range of important related topics. These include an exploration of the deterrence equation as it applies to Australian defence in the alliance context, Australia as an industrial anchor for defence production, the littoral context, the concept of operation for the use of autonomous systems and the path to best leveraging their effects.

Laird also engaged with ANU colleagues like Dr Andrew Carr, Professor Stephan Frühling and Jennifer Parker. Laird tells us Carr argued Australia was not required to 'do deterrence in the Cold War' and therefore needs to think afresh about what this means when it comes to a new submarine capability.

Professor Stephan Frühling suggested Australia's defence contribution was largely about 'horizontal escalation', working with allies and regional partners to bolster their national defences and encouraging greater European engagement in Indo-Pacific security arrangements.

Jennifer Parker spoke about the evolving threat environments in the littorals, with many lessons drawn from operations in the Black Sea, Red Sea and Philippines Sea. Then there were others such as Gregor Ferguson and Dr Malcolm Davis (focused on space and cyber).

Included in the appendices is the full text of Lieutenant General Simon Stuart's presentation to Land Forces 24, in which he addresses 'the human face of battle and the state of the Army profession' – a topic to be approached 'with a great deal of humility, and respect for war's unpredictable nature.' This historically-infused review of the place of land forces in the defence of Australia and its interests is well worth reading closely.

Another appendix, by John Blackburn and Anne Borzycki, addresses the impact of the Australian political system on national security. Their reflection on defence policy decisions in recent decades notes their concerns that 'national security continues to be conducted through a somewhat narrow military platform lens.' They note that the 'dismal and costly' consequences of Australia's political decisions to get involved in wars in Vietnam, Iraq and Afghanistan, let alone the acquisition of nuclear propulsion submarines absent a wide debate. It is not surprising; therefore, they argue, that 'Australia lacks a national risk assessment, [and] a national security strategy'. The need for a national security strategy is as great as ever.

This book provides grist for the mill for advocates of a holistic considered national security strategy that makes clear what is required, the urgency of action, the expenses likely to be incurred and the reason why the electorate should be persuaded.

By John Blaxland

John Blaxland is Professor of International Security and Intelligence Studies and Director of the Australian National University's North America Liaison Office, based at the Australian Embassy in Washington DC.

CONTENTS

INTRODUCTION

The modernization of the ADF is a case study in the clash between force design for an envisaged future force and the need to enhance the force in being to deal with the world as it is.

The challenge for any force planner is to combine a projection of desired core platforms, connectivity technologies, the nature of the adversaries to be dealt with by such a force and to do so from the perspective of what we believe of the future envisaged from the present.

It has always been difficult, but it may be well be the case that future force planning built around projections of the platforms of the future has outlived its day and is perhaps counterproductive.

What good is a thirty-year shipbuilding program when we are entering a period which will be significantly reshaped by maritime autonomous systems?

And are we preparing for World War III or are we shaping a strategy to sort through how to deal with the gray zone conflicts and local wars characterizing the multi-polar authoritarian world?

This difficulty of force planning is enhanced by how today's ready forces need to modernize to deal with current operations.

Or put another way, closing the gaps for the fight tonite force generates change which simply is not captured by future force planning built around iterative platform replacement. You are not going to capture the nature of future air warfare by designing a so-called sixth-generation fighter to replace a fifth-generation fighter.

We have entered the world of the kill web where the evolution of warfare is being shaped by the payloads which can be configured and connected to create the desired effects in a combat space. We are building combat clusters rather than platform identified task forces. We are building maritime combat clusters not destroyer task forces. And this approach decisively challenges a platform defined future force planning approach.

The Australian government has identified a number of characteristics of the future force. The problem is that for the current force to be effective, lethal and survivable, it needs to upgrade its force on the fly.

And this requires more attention to rapid modernization than has been pursued in the past two decades.

For example, Jennifer Parker, the noted Australian naval strategist, argued this about the implications of the changing littoral environment for capital ships:

I had a chance to follow up with Jennifer Parker on her excellent presentation at the 26 September 2024 Sir Richard Williams seminar which focused on the evolving threat environments in the littorals and insights to be gained from operations in the Black Sea, Red Sea and the Philippine's Sea.

She argued in that presentation that new capabilities, notably USVs and UAVs used by the Ukrainians and the Houthis posed new challenges to capital ships in the littorals. And that capital ships clearly can still be effective but ongoing modernization of their defensive systems in the new context on an ongoing basis was critical.

In our discussion, she underscored that the threat from land systems was of enhanced range, and new threats posed by unmanned maritime systems introduced additional threats to capital ships as well.

What this meant for her was the absolute importance of ongoing modernization of the combat systems aboard surface ships. Rather than viewing updates as occurring in long periods of block upgrades, there needs to be an ability to weave in upgrades based on the rapid evolution of offensive threat systems from various operational theaters.

It is crucial to have very credible threat information from a diversity of deployments by the Royal Australian Navy and its allies, and to be able to weave that information into ongoing upgrade efforts of combat systems.

And modern airpower is being generated by software upgradable systems such as the F-35 which need regular upgrades to keep pace with the rapidly changing combat environment.

The need to upgrade more rapidly dips into the costs of funding the future force or alternatively if you starve the current force to pay for a future force you threaten its viability.

But as the September 26, 2024 Sir Richard Williams seminar focused on the ready force, a further challenge for future force planning was laid bare. Virtually every major speaker who spoke about how to ramp up the capabilities of the ready force highlighted the salience and importance of incorporating autonomous systems within the force.

But how to do so is a major challenge in part because these systems do not follow the development and operational path of legacy future force building models. They are payloads more than they are platforms, and as with all payloads their utility is determined by actual warfighters, not think tank strategists and force planners.

This character of the evolution of the ready force incorporating autonomous systems was highlighted by the former chief of the Australian Navy Vice Admiral Tim Barrett in an interview I did with him.

This is how Barrett put it in the interview:

The reality is that for autonomous systems to come into the current force, they need to be well practiced at the operational end – promotion of their adoption is a behavioral piece. New systems need to be in the hands of warfighters to ensure that these systems make the current force more agile and take actions that are effective in their application.

Operational success is still about the application of force in the mind of the person responsible for delivering it. It's about forcing them to think of operational success by whatever means they have available to them, and then having the courage to take those actions.

You are not necessarily after disruptive change in process, but disruption in the effect. In some cases you don't want disruption in the efficiency of the process of operations. But, you want to be able to cause a disruption that has an effect on your adversary.

With regard to the Ghost Shark, to fully achieve its potential, it has to quickly enter the operational world of those who are managing the underwater warfare space throughout the regions of our interests.

To be effective as a disruptive technology, it will need to contribute to the operational effects being sought by those managing the undersea domain; in tactical terms this means it has to be of benefit to those managing the water column. It could generate strategic consequences but not simply because of its technology but in the way this it is used to produce disruptive operational effects.

A successful water space management process is key to being able to determine where your adversary is, or, more importantly, where it isn't, so that you can put the right forces in the right place.

Bureaucracies don't necessarily think like that. Operators absolutely do so, because it's their day-to-day business, and they're in the practice of only putting in harm's way those things that need to be there to affect a disruption to the enemy's operations.

The disruptive effects that a Ghost Shark can produce should be determined by those who actively manage the battle space, the undersea battle space, rather than someone who's programming from afar and doing so in complete isolation from the rest of the water space management concepts of operations.

In other words, the Australians face a challenge common to the United States and its allies: How is the changing nature of more rapid upgrades in the ready force affect the practice of future force planning?

There is probably no starker reminder of how things have changed than watching the Israelis execute a strategy in their seven front war. They have combined fifth generation aircraft with exploding pagers and a variety of new combat clusters to deal with a deadly threat generated by Iran through its complex web of warfighting.

Does anyone really think that force planners sitting down a decade ago envisaged the force engaged today by the Israelis?

Rather the Israelis have modified in a variety of innovative ways the ready force, adding some new core platforms, but focused on a kill web enabled force to deal with a range of threats.

We are at a turning point in our military modernization strategies.

Are we paying for a future force while being compromised in the conflicts of today?

Now let me turn to the book itself. The main body of the book is my report on the 26 September 2024 Sir Richard Willians Foundation seminar which focused on the ready force and the challenges to more rapid modernization.

In addition, I have included interviews which I conducted during the September-October visit to Australia. These interviews complement the report and provide further depth on the Australian strategic community's thinking about the way ahead for the ADF.

Finally, I have added five appendices to the book to provide the reader with additional insight with regard to the challenges facing ADF modernization.

Appendix 1 provides the seminar description and program.

Appendix 2 provides further insight into the perspectives of the Chief of the Australian Army, Lt. General Stuart Simon. The speech he gave at the land forces conference in September 2024 in Melbourne provided a comprehensive overview of the way ahead from his perspective.

Appendix 3 is a seminal essay by John Blackburn and Anne Borzycki. Their essay addresses the challenges which the political system in Australia poses to having an effective defence policy. Australia clearly needs a national defence and security strategy and not just the AUKUS totem.

This essay was originally written for Murielle Delaporte's publication *Operationnels* and a lead article in her journal publication prepared for the Eurosatory 2024 Conference, held in Paris in June 2024. John is a member of the editorial board of the journal and wrote this essay along with his colleague Anne Borzycki for that publication. We thank the editor and the authors for permission to include an edited version in this book.[1]

Appendix 4 provides some additional analyses which were published on our website *defense.info* during 2024. The first is by the Paris-based Pierre Tran on the Australian submarine decision. The second and third are by the noted Australian journalist and analyst Greg Ferguson.

The first of Ferguson's analyses looks at a proposed Royal Australian Navy response to the opportunity to work with autonomous systems. The second examines how the ADF is shaping its long-range fires acquisition policy.

In short, this book provides insight into the state of play concerning thinking about the modernization of the ADF as a ready force as seen in the second half of 2024.

PREFACE TO THE 26 SEPTEMBER 2024 SEMINAR

Prior to the 26 September 2024 Sir Richard Williams Foundation seminar which focused on shaping a way ahead for the ready force, I spoke with John Conway to get his perspective on the focus of the seminar.

With four decades experience in the business of air combat across operations and industry, John is identified this way on the foundation website:

John is the Managing Director of Felix, an independent company providing specialist capability development and operational analysis services to Defence since 2017. He was previously a business development and strategy executive with Raytheon Australia specialising in air combat integration, electronic warfare, advanced weapons systems, test and training ranges, and integrated air and missile defence.

John retired from the Royal Air Force as a Group Captain in 2010 having served 24 years in a number flying, staff and senior command roles. His operational experience on F4 Phantom and Tornado F3 aircraft included Cold War Europe, the South Atlantic, the Balkans, and the Middle East. He commanded the United Kingdom's largest Permanent Joint Operating Base at RAF Akrotiri in Cyprus between 2005 and 2008 enabling the airbridge into Iraq and Afghanistan, and supporting strategic ISR operations in the eastern Mediterranean.

Conway underscored that the Sir Richard Williams Foundation seminars had most recently focused on the strategic redesign of the government for the ADF. But the future is not yet here, and the question is how lethal and survivable and effective is the ADF now in its operations? How can it be more effective in the near term? And how are the ADF and defence industry focused on doing so?

"For defence leadership, there is a very complex challenge of balancing investments in today's battles and the future ones. But to understand the challenges for the ready force we need to get outside of the Canberra environment and understand how operational commanders and the ready force are addressing current challenges and shaping a practical way ahead.

"We have threats squeezing the defence force. We have a challenging fiscal environment. Despite starting from a less-than-ideal situation, how do we operate more effectively in the here and now while keeping an eye on the future? How do we get everybody to work together to create outcomes greater than the sum of the parts?"

He then discussed the late Cold War period where the West understood that it was at a disadvantage with the Soviet Union in a number of areas but worked on building its own asymmetric advantages as part of a credible deterrent structure. He argued that we need the same attitude and approach for today's world as well.

By focusing on our competitive advantages and enhancing them, one can build a pathway to reinforce the ready force and build for the future.

He identified a number of areas for particular attention where activities today can shape the development of the future force.

The first is the sophisticated exploitation of the electromagnetic spectrum in both defensive and offensive counter air missions. Here he discussed the importance of enhancing the resilience and redundancy in our systems and honing a kill web approach which makes the denial of core ADF electromagnetic capabilities very difficult and hence enhancing the overall deterrent effect.

The second was the question of shaping credible ways ahead to build combat mass. Here the focus is upon building and operating within the ready force various types of autonomous collaborative systems and working them in as key enablers of the payloads necessary for evolving a significantly enhanced integrated force.

The third is strengthening the enablers of a kill web, notably in terms of C2 and distributed ISR and the ability to perform effective counter-ISR against the adversary. Central to this is an enterprise approach to mission data management and reprogramming for the integrated force. Work is being done to enable the ready force now and how it is being done is shaping the future force as well.

He concluded with a very important core point about leveraging and building on the ready force and our lessons learned in the Cold War as we build the future force.

"I think going back to some of the traditional ways that we've done things in the Cold War and making them more contemporary by the exploitation of new radio frequency technologies and putting the money into the back end of the systems, through the networks, the comms architecture, the mission data environment, and start bringing tempo as well to the way we operate.

"It sounds a little bit more like an evolution rather than a revolution. And I think that's fair. We don't need to destroy everything that we've learned in the past. We just need to keep investing in that next stage and accept that the future probably is a smaller available workforce, and this means we need to effectively exploit technology such as autonomous systems for an enhanced ready force."

1

OVERVIEW TO THE 26 SEPTEMBER 2024 SIR RICHARD WILLIAMS FOUNDATION SEMINAR

The most recent seminar of The Sir Richard Williams Foundation was entitled: "Enhancing and Accelerating the Integrated Force: An Operational Perspective." This seminar focused on how the force in being or the ready force was focused on improving its capability and its survivability and lethality for the fight tonight. For such a focus, one needs a sense of urgency for how to improve the force in the short to mid-term, rather than over emphasizing the long term future and future force structure planning.

This seminar begins the third phase of The Sir Richard Williams Foundation seminars since I have been writing the seminar reports. When I first came in 2014, the seminar focused on "Air Combat Operations: 2025 and Beyond." Well, 2025 is here. And 2014 was as well the first year that the Marines began their six month deployments to Australia or MRF-D showed up in Darwin.

This first phase focused on modernization of the RAAF, in terms of a crafting a fifth-generation air force and the impact and intersection of this effort with the overall modernization of the joint force or the creation of a fifth-generation enabled force. This focus continued until the pandemic hit in 2020. The work conducted in the seminars during this period is summarized in detail in my book published in December 2020 entitled *Joint By Design: The Evolution of Australian Defence Strategy*.

The second phase has focused upon the strategic redesigns envisaged for the ADF first by the Liberal government and then by the current Labour government. Here the focus was upon discussing and analyzing the way ahead for long-range force planning and strategic redesign of the force.

We focused on these topics in the seminars which began after the pandemic lifted and through the first seminar of this year. I have summarized the work of this phase in two books, *Australian Defence and Deterrence: A 2023 Update* published last year and *Australian Defence and Deterrence: A 2024 Update* published in 2024.

Now we are launching a third phase focused on how the ADF is enhancing the force in being from an operational perspective.

A rule of thumb is that a military has 80% of the force it will have in a decade, so the challenge is always adapting the force in being while new platforms and systems are added 10 to 20 years into the future.

There were several themes discussed during the seminar.

The seminar began with a presentation by Peter Jennings who focused on the political dynamics within the three AUKUS countries and how those dynamics might affect budgets, operational demands and force structure choices in the five year period ahead. Quite obviously the fight tonite force is clearly affected by operational demands and the ability and willingness of governments to provide sufficient funds for near term operational demands as well as funds for longer term force modernization.

A key challenge as well is the need to broaden the consideration of how defence notably in a multi-polar authoritarian world challenging the rules-based order has become a much broader consideration for the Australian nation than simply what the ADF can deliver. Notably, the information society is a bedrock for modern Australia, and enabling, protecting and securing the information tools of a modern society is a whole of Australia problem, not just an ADF issue.

Three speakers especially noted this aspect of the way head. Chris McInnes, Executive Director of the Airpower Institute, underscored that the central role of aviation within Australian society requires rethinking the whole of aviation enablement challenge, rather than simply focusing on military airpower. There is a much broader eco-system for air power in defence of Australia to be considered.

Space systems clearly fall into the same class of challenges, whereby space for Australia is a whole of nation domain, not just a warfighting domain that requires only ADF-competence. Nick Leake, Head of Satellite and Space Systems at Optus, has made this point repeatedly in his past presentations, and did so as a member of the panel of industrial representatives.

Dr. Malcolm Davis of the Australian Strategic Policy Institute has underscored in his extensive work on space and in his presentation, the need to ramp up the investment and work of the Australian government more generally on the space domain because of its basic significance for the viability of the information society and economy of Australia even beyond narrowly considered defence considerations.

Throughout the day, there were discussions on the need to modernize continuously the enablers for joint operations, C2, ISR, Counter-ISR, or what my colleague Ed Timperlake calls the essentials of prevailing in "tron warfare."[1]

The Chief of Joint Capabilities, LTGEN Susan Coyle argued that this area is indeed the high ground of military capability. And I would add the further challenge is that this is a software dependent world which is evolving rapidly and the capabilities needed are ongoing ones and simply are not long range force structure platform planning targets.

The Deputy Director General-Signals Intelligence and Effects from the Australian Signals Directorate certainly reinforced Coyle's emphasis on the role of tron warfare and intelligence as an enabler for the joint force and in this way it becomes a force which can operate as an integrated one.

The most targeted discussion of the force in building and the approach to change from an urgent shorter-term perspective was provided by a core panel. This panel was headed by AIRMSHL (Retd.) Darren Goldie and the two members of the panel were the Air Commander Australia, AVM Glen Braz, and the Commander of the Australian Fleet, RADM Christopher Smith.

Commander Australia, AVM Glen Braz, on the right, and the Commander of the Australian Fleet, RADM Christopher Smith, on the left, participating in the seminar.

Both focused on priority modernization efforts to work force distribution for the air and naval forces to be more lethal and survivable.

We benefited as well from a video presentation from General Kevin Schneider, Commander Pacific Air Force, and in person presentation by AVM Mark Flewin, Air Officer Commanding 1 Group Royal Air Force. The first discussed the challenge of modernizing the force in being from an American point of view and the second from a British point of view. And with the AUKUS emphasis, there is growing convergence on force integration issues with Australia.

Jen Parker of the National Security College discussed the evolving maritime environment and its challenges, notably the changing littoral operational environment and the appearance of new technologies. The Chief of Air Force, AIRMRSHL Chappell, discussed the importance for the RAAF being able to operate its force in depth, in terms of space, time and posture.

Throughout the day the growing significance of autonomous air and maritime systems was mentioned as a key element of the way ahead for near term modernization of the force, and this along with the tron warfare evolving capabilities were obviously key building blocks for the future. And this ultimately is the challenge: How does enabling the force in being, the fight tonite force, in the near to midterm mate with long term visions?

This is a major challenge for industry supporting defence whether those are defence companies or commercial companies. How does industry make the investments in defence modernization which upgrades the force in being to fight tonite?

A key issue involved in all of this is how to enable the acquisition process to fill the gaps which the current force must meet to become more survivable and lethal for the fight tonite rather than prioritizing 10 year to come platforms?

2

THE CHALLENGE FOR DEFENCE READINESS: THE IMPACT OF POLITICS

If one is focused on how the force in being can be a more ready force, one will generally look in vain in the political class for a keen focus on this challenge.

And this is true for all of the AUKUS partners.

It is not difficult to see why.

A ready force needs supplies, munition stockpiles, reliable energy supplies, food stocks, logistics capability and an ability to mobilize civilian and reserve military manpower. All of which cuts into spending for social programs, envisioned transitions to the green economy and supporting whatever party is in power's pet rocks on defence projects.

It is also the systemic bias towards short-termism in defence thinking as well as the desire of new governments to craft alternative defence futures with weapons that are not here and now. This is true across the board for the three AUKUS countries.

As Air Vice-Marshal John Blackburn AO (Retd) and Group Captain Anne Borzycki (Retd) wrote in their recent essay on the impact of politics on defence (See Appendix 2):

Short-termism, or 'quick win' thinking, is deeply embedded in the

Australian political culture; collectively we tend to focus on today and largely on our personal needs, not on future interacting and cascading risks that will impact our whole society.

Thirty years of relative prosperity in Australia, fuelled by lower trade barriers, privatisation, and deregulation, have increased our productivity and wealth, providing the resources necessary to address the challenges we confront today if we choose to act.

However, many of these challenges are themselves a result of globalisation, e.g. our extensive reliance on overseas supply chains for critical goods, leaving our nation vulnerable. Whilst the lower cost of goods has had economic and standard of living benefits, there is a very high price to being 'cheap' in a crisis.[1]

At the September 26, 2024 Sir Richard Williams seminar, Peter Jennings took on the task of looking at the politics of the AUKUS nations and the impact of those domestic politics on defence spending and preparedness. He focused on looking ahead in the second half of the decade and what politics in AUKUS nations during that period of time might mean for defence.

The core point he made, and a key one it is, was to emphasize that "defence policymaking is a generation of politics by other means." And quite obviously the priorities of the party in power will decisively affect defence choices and policy making.

Jennings underscored that the 36-month election cycle in Australia clearly impacts on the time-frame for defence decision making.

He reminded the audience that although politics and economics are decisive for defence, too often professional conferences on defence simply ignore this reality.

He underscored: *What I find slightly unusual is that in professional defence conferences, there's a tendency to put that to one side, to pretend that it doesn't really exist, to make presentations that would argue that the shape of defence policy and spending priorities are things which kind of happen as a result of mutual great minds thinking deep thoughts.*

And I want to make the case this morning that in fact, politics has far more influence over the shape of defence policy than such a point of view would consider.

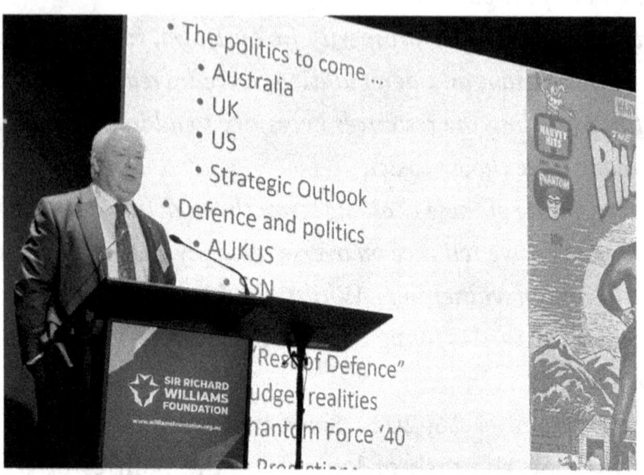

Peter Jennings presenting to the Sir Richard Williams Foundation, September 26, 2024.

I would like to reinforce this key point of Jennings.

Later this year, I am publishing a book of essays by Dr. Harald Malmgren, one of the most distinguished political economists in our lifetime and who served several Presidents and was a major shaper of U.S. trade policy in government and the private sector.

I would bet most people who attended the seminar never heard of him, but perhaps they might have heard of his dynamic daughter Dr. Pippa Malmgren, who was recently in Australia.

I have worked with Harald off and on since 1980 and our work has been at the intersection of the Venn diagram between defence and economics, each of us helping the other to be up to date on these two domains, because these two domains decisively affect one another.

But this is a struggle because Inside the Beltway makes defence strategic decisions often with no consideration to cost or impact on the American economy, something which has decisively affected American power and accelerated its decline.

Jennings assessed each of the AUKUS partner's political situations over the next few years and how they might impact on defence. The implications of his assessment were pretty stark: the three countries are not necessarily on a convergent path to strengthen collective defence or to provide the defence spending necessary for the ready force.

When he turned to considering the future of AUKUS he made some hard-hitting judgments.

Phantom Force '40

- "integrated force" is <u>never going to happen.</u>
  - $$$
  - AUKUS SSN or the ADF
  - Ignores strategic reality driving alliance priority over ADF integration
  - Not a political priority
  - Damages the current ADF.

He argued: *What disturbs me is that there is no federal, no industrial, no state, no union advocacy for AUKUS. There is no clarity around the East Coast port and no further progress on the development of a uranium waste storage facility.*

This is a project which, right now, in public, has fewer and fewer friends, and I reflect on how the Shortfin Barracuda project had fewer and fewer friends until it was terminated.

There's no money, virtually no industry advocacy, no sense of urgent goals, no war fighting priority to get equipment to into the hands of our war fighters on an urgent basis. We have what a friend of mine in industry described as a series of science projects...

The good news is that I think there's been great progress, useful progress, on regional cooperation, gaining traction on the guided weapons enterprise, on the shipbuilding enterprise in the West of Australia.

The negative side of the story is that the submarine is eating the budget...And that leads me to conclude is that the ADF is less ready, now, less capable now in 2024 than it was in 2022.

That brings me to Phantom Force 40, The Future Force, the Integrated Force.

Let me make the proposition to the integrated force never going to happen. And that's because there's not the money for it. It's because right now, the policy presentation of government is putting is that you can have the SSNs, or you can have the ADF, but you can't have both.

The integrated force concept also ignores the reality that what's driving the big movers of strategic thinking in Australia right now are Alliance priorities.

It's about how we position ourselves to deal with the challenge which China is offering now, rather than over ADF integration.

I think we should say farewell to Phantom Force 2040, because I just don't think it's going to happen...

I'm predicting 2026 as a use it or lose it moment. What I mean by that is to say it took about five years to get to the point where we concluded that the French submarine deal was not going to work and we needed to move away from it.

I think it's going to take about five years to bring government to the point where it is going to say do we really want to do this, once they understand the cost of the complexity involved. Not saying they should walk. I'm just saying that whichever government is in power in 2026, a decision will have to be made about the submarine's future.

3
BUILDING COMBAT MASS: A NAVY PERSPECTIVE

When you are a medium-sized force and not adding a lot of manned platforms in the next five years, how do you enhance combat mass?

Or put another way how does the ADF as a ready force create mass effects with what they must fight with tonight and what they can plus up with in the short to mid-term?

This was the focus of the panel chaired by AIRMSHL (Retd) Darren Goldie with participation from the commander of the Royal Australian Navy, RADM Christopher Smith, and the commander of the Royal Australian Air Force, AVM Glen Braz.

Let us start with RADM Smith's remarks and then unpack them.

Considerations of agility are paramount to our force structure. But how is this achieved framed against the scope and the pace of today's military challenges, while preserving the flexibility to respond to contingencies of climate change and mass migration?

How do we generate the necessary mass to deliver on this broad remit of diplomatic constabulary and warfighting missions?

Generating mass will continue to challenge us, but I will attempt to articulate specific methods I think can assist. I'll focus particularly on generating mass by increasing survivability of our forces, generating mass through partnerships, both with industry and in our region, and finally, generating mass by developing multistage force structures that add scalability, resilience and agility relative to our traditional conceptualisation of mass. Today, the challenge we face pairs near persistent, wide area surveillance capabilities with advanced, long range precision strike.

Moreover, the volume and pace of the PRC's naval program has altered the landscape in which we must conceive of mass. Collectively, these challenges require us to reappraise our methods of confronting that challenge.

The first way in which today's RAN is able to build combat mass and depth is by incorporating distribution forces operating across the maritime domain aim to defeat our adversary's ability to find, fix, track, target and engage Australian forces.

This methodology seeks to distribute our forces in less detectable, less targetable means that present no clear center of gravity to the opposition. This approach eschews optimizing defeat of the adversary's missiles in lieu of defeating their ability to effectively employ them.

Distributed forces are also able to match their firepower across a wider area compared to denser concentrations.

In this way, our forces become more survivable and better able to safeguard our maritime communications and trade. Echoing the wisdom of the classic theorists, distributed maritime operations places emphasis on massing effects vice platforms to generate the necessary depth of lethal force at the decisive point.

Distribution as a core concept of our operations therefore seeks to manage a defensive problem while seizing an offensive opportunity.

But there is tension between greater distribution and effective C2. Distributed forces will need to be supported by scalable and flexible C2, elements able to operate, either ashore or afloat and remain connected by resilient, low signature, redundant data networks that can withstand the contest for spectrum.

To support our survivability and decision advantage, we must dominate the various spectra in which our forces operate, employing techniques of deception and maneuver to install doubt in the mind of our adversaries.

The contest for spectrum will impact the wider contest for decision making advantage critical to our ability to dictate the tempo of conflict.

Together, these concepts will enable the RAN to achieve electromagnetic mass simultaneously flooding the spectrum and manipulating the pace of decision making of our adversaries.

Decision superiority and the ability to dictate the scale and tempo of operations will thus be generated through distribution and manipulation of the perception of mass that we present to our adversary.

This brings me to the second point on how we will deliver mass. The concepts I've outlined will require leveraging our critical partnerships across industry and across borders.

The pace and scale of change is perhaps nowhere more obvious than in the conflict between Russia and Ukraine. Of particular note is the ability of Ukrainian forces to generate a sea denial capability against the Black Sea Fleet through employment of remote and autonomous systems delivered by industry partners to fill a critical capability gap for Ukraine.

The full impact of what these developments may mean for potential great war conflict in our own region may not be fully known, yet it appears there are opportunities available for their employment as a complicating feature of the contested physical and informational battle space.

The potential of these systems to raise the noise floor for our opposition may go well beyond their presence in the physical domain to what signatures can be synthetically manipulated and what synergy may be found with loitering munitions and decoys.

The full realization of these capabilities to generate electromagnetic mass will necessitate the type of technology and information sharing with our partners that AUKUS can deliver and will be reliant on our continued engagement with domestic industry partners to push the boundaries of what is viable.

Finally, pairing the two points above, the RAN will seek to generate mass by reshaping its multistage system that leverages the potential of modern, uncrewed systems and spectrum manipulation with the long range understood benefits of distributed forces in the maritime domain.

A multistage system can be understood as a combat force that maximizes the versatility, adaptability, survivability and controllability of multiple layers of the fleet and maritime forces.

The combat power of a stage system is simultaneously concentrated and distributed to generate mass in the battle space that commanders will leverage to dictate and scale the tempo of combat.

Importantly, a multistage system requires all components to be defeated in order to negate its combat power. This increases the resilience of the systems, degradation and attrition.

The next evolution in a multistage combat systems will almost certainly involve increased reach, persistence and agility by inclusion of remote and autonomous systems that are able to pair with or operate independently from their associated crewed platforms.

Future multistage systems will incorporate adaptability and interchangeability that goes beyond the power of carrier centric forces, as the diminished size and the crewing requirements make this capability available and affordable additions to even frigate and destroyer sized platforms.

But more than just uncrewed systems, future multistage forces will leverage the joint capabilities of distributed land and air assets throughout the maritime environment. The ability of future ground and air forces to contribute essential kinetic, non-kinetic effects to multi domain strike missions in the maritime environment will be essential to improve the economy of effort and generate sufficient precision and firepower at the decisive points.

The conception of mass in modern maritime combat demands clarity in the development of our doctrine and concepts. As past examples have shown us, challenges Australia faces today are not unique, and our thinking around combat, mass and depth should be informed by previous incarnations of this conundrum. So we must again today, reconceive our thinking around mass.

Our fleet has embarked on this journey, but full clarity around the destination remains ambiguous.

Concepts of distribution and spectrum manipulation will support our requirement for decision superiority. Uncrewed systems will add complexity to the adversary's situational awareness while simultaneously refining our own.

Pairing these ideas as a network of scalable and adaptable teams to form multistage systems will build necessary depth, versatility and resilience.

Yet the final product remains to be fully conceived, articulated and engineered to success.

Engagement with industry and coalition partners will remain pivotal to delivering this end state. Along the way, we will require no small amount of innovation and an appetite to deal with the concomitant risk.

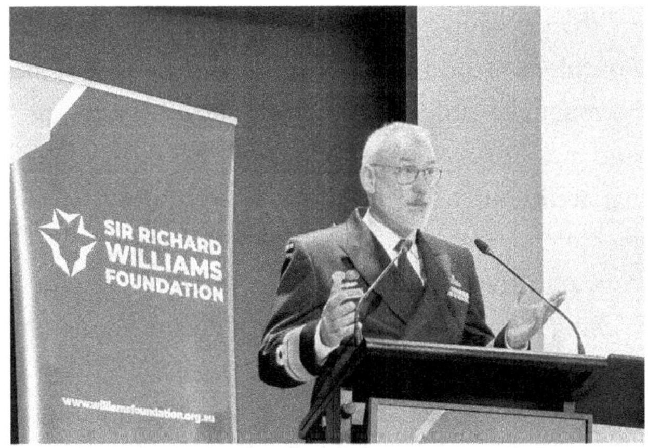

RADM Smith speaking to the Sir Richard Williams Foundation
September 26, 2024

What makes this presentation so interesting is that it combines ways to enhance the current force going forward with the end state which he sees as necessary for the Royal Australian Navy. In the near to mid-term, one needs to enhance the ability of the fleet to be augmented with the addition of uncrewed systems but to do so by having significant creativity in generating combat clusters afloat which can leverage ground and air capabilities as well supporting distributed fleet operations.

While not talking directly about ship building his emphasis on mass effect is upon what he calls a multi-stage system or what I would call a maritime kill web. The idea is simply that through ISR, C2 and Counter ISR, the maritime maneuver force is able to mask, and to integrate with joint or coalition forces to deliver the mass effects necessary to confuse, mis-direct and defeat adversary forces.

It is really about building kill web maneuver forces. In a recent interview I did with LtGen (Retired) General Heckl, the recent retired head of the USCM Combat Development Command underscored that what Force Design for the USMC is really all about was creation of a maneuver force able to operate against a force with improving ISR and significant long-range precision strike.

And in that interview, we focused on what I see as a major unsolved problem – the need to use autonomous systems now as part of the maneuver force and how to begin solving the ability of manned and autonomous systems to work effectively in crafting the required capabilities of a successful maneuver force.

One problem he highlighted was the following:

"How we effectively communicate with autonomous systems is the key to using them effectively. We don't want rogue robots in the battlespace. And bandwidth is a key challenge.

"How do I work with multiple autonomous and manned systems? How do I communicate? How do I exercise fire control? And how do I provide the kind of interactive support and guidance with the Ground Combat element as it works the kind of offensive-defensive maneuver required in the conditions of being threatened by adversarial long-range fires and capable surveillance systems?"[1]

This is a key element for working the kind of multistage system Smith is talking about. And this is not a long-range problem. It is part of the challenge in the next 2-5 years and as it is worked, the scope and nature of the future force will change significantly in ways no future force planner can now accurately predict.

This raises what I think is a key conundrum: If the force structure in being is modernized through the digital enablers for an integrated force –- C2, ISR, and Counter-ISR – and autonomous systems which are also software upgradeable payloads, how then do you know what exquisite platforms to buy for the future?

4

BUILDING COMBAT MASS: AN AIR FORCE PERSPECTIVE

The panel on combat mass was chaired by the former Air Commander Australia, AIRMSHL (Retd) Darren Goldie, who provided an overview to the discussion.

AIRMSHL (Retd) Darren Goldie introduced the panel with a clear overview regarding the challenge.

This is how introduced the subject:

The definition of combat mass is widely agreed as overwhelming combat power at the decisive place and time. And it's worth remembering that this is about massing combat effects, rather than massing combat forces.

It's worth remembering that this concept is not new. To quote Clausewitz ***On War****, the center of gravity is where mass is concentrated. We hear a lot of negativity in Canberra: There's not enough money, there's not enough people, there's not enough strategy, there's not enough drones, there's not enough everything.*

We need to uplift this conversation and talk about the great things that are happening in the fighting parts of our force. Between the two speakers, they command 40% of the Australian Defence Force, and it's a rare opportunity to have the two operational commanders in front of you and talking about how they are thinking through the most complex problems facing the operational force.

AIRMSHL (Retd) Darren Goldie charing a panel at the Sir Richard Williams Foundation Seminar September 26, 2024.

After the presentation by Admiral Smith, the current Air Commander Australia, AVM Glen Braz, then spoke. Let me start with his presentation and then unpack some of his thoughts to discuss the challenge of ramping up the capabilities of the ready force.

AVM Braz highlighted a number of key themes in his presentation:

Building combat mass is front and center of all of our efforts across Air Command and Air Force at the moment, and I like to think of it across a few layers. And those layers are broadly the technology layer, the human layer, the operational layer, and finally, and perhaps most importantly for us in this context, is the organizational layer.

From a technology layer perspective, there is no doubt we need the enhancements of proliferated UAS and UAV systems that are collaborative in human and machine teaming, and we are working to build the basic steps of bringing that into service as soon as we can, pending the outcomes of the good work across MQ 28 and other similar systems.

I do need off-board weapons systems, and I do need off board sensor systems as part of a mass force.

AVM Braz, Air Commander Australia, speaking at the Sir Richard Williams Foundation seminar on September 26, 2024.

We need to be able to concentrate massive force at the time and place of our choosing. And it's going to be a relative measure, one that will be undoubtedly constrained by the physics of our force which remains small.

I absolutely need UAS for combat effects, but I also need them for survivable distributed logistics...

In developing agile operations concepts and to orchestrate the operation of the force, I need to understand what's going on across the force. Situational awareness and resilience of communications are fundamental to how we achieve the tactical and operational effects that we need... to be able to synchronize and harmonize the mass effect that we need at the right time and place.

Cost effectiveness in terms of affordable and attritable systems need to be part of that concept.

We are outstanding at building long duration projects with exquisite platforms... But they're not perfect, and they need ongoing investment. They need ongoing sustainment and upgrade to remain relevant, because even the major systems like F-35 need updates, and they need to be brought up to the latest standard.

It's no different for the enhanced force in being, where we need to embrace the concept of more consumable capabilities that are more quickly adapted into the battle space and that can give us affordable and rapid mass if and when we need it.

And I'll talk more with regard to how we get after doing some of that in terms of manufacturing and adaptation that is going to be key to generating and sustaining mass. And I have already mentioned that the ongoing ability to remain aware of the battle space where our own forces are in the distributed model will be fundamental to what we do.

If I look at the human layer, we have scaled up pilot training and the number of graduates. I'm actually at the point where I have to throttle the number of graduates for air force through the pilot training system at the moment, because I need to be able to absorb them downstream, and I need to sustain and operate them downstream as part of a broader force.

And while I can grow elements of the workforce as directed by the national defence strategy, I need to buy and grow all parts of the system with a degree of proportion or the system will be misshapen and will fail as a result.

So pilot training is a success story for us, but equally, we've ramped up our ability to scale initial military training and initial employment training across the air domain. There's workforce allocations available across the department and in the air domain and I'm confident that we can succeed. We are attracting more people than ever. People want to come and join our Air Force and the ADF writ large. And I want to leverage that, and I would absolutely welcome the apportionment to the air domain to let me to continue to grow in the right shape and in the right proportion...

We're working very hard to get better at how we apportion and appropriate relevant or minimum viable skills to the workforce...

The human layer is about our mindset. For example, with regard to the Combat Support Group, we are changing the mindset of that organization to be one that acknowledges that everyone is in the fight, not one wing over another, not a garrison wing versus a deployable wing, but the whole of Combat Support Group is adapting to the mindset that they will be needed to go forward to generate the mass and the resilience and the depth in the force to operate across the bases and places that we need them to function...

When you're a small force with outstanding capabilities, you need the absolute most from your people, and you want to generate mass in time and place of your choosing, then having the ability to surge key parts of that workforce will be fundamental...

The design of how we would cluster and disaggregate and reaggregate is something that's a core activity, mostly led by the Air Warfare Center, and I'll come back to them in the organizational mindset in a moment, but they are working on the ideas of deception and and dispersal and how to maximize how to be survivable, including hardening and other elements that are very important to us.

Fundamental to all of this will be interoperability with our key allies and partners, and while we can generate interesting agile operations mindsets, unless they are well integrated and nested with our partners then our ability to collaboratively generate mass will suffer. I'm working very closely with U.S. Pacific Air Forces as one example to enhance our interoperability across the C2 and across the weapon system that is the air operations center.

Finally, let me discuss the organizational layer. I want to see a pivot in our defence organization that is underpinned by what has been often mentioned this morning, namely a sense of urgency... I would love to see that permeate more broadly across the defence department.

The pivot that we need to take is one that moves away from the mindset of long term, very large, long duration, low risk, in some ways, projects to deliver exquisite capabilities to one that is more adept, structured, organized, and has the risk tolerance to take on short term consumable capabilities that can rapidly and with relevance uplift the force in being.

That might take some time and will take commitment from across the organization to achieve that sense of urgency. [For the RAAF, one way to do so] is to leverage my Air Warfare Center which is a crucible for creativity and ideas and where we can bring new technologies into the force quickly. We can experiment with them. We can apply them to great training scenarios, and we can inform a rapid capability cycle inside the system in ways that will be tangible and meaningful for the enhanced force in being and help us generate mass....

Let me unpack some of these thoughts and make some comments.

As Air Commander Australia, Braz is charged with being ready for the fight tonight. He must be focused on a sense of urgency. One area where a lack of a sense of urgency is evident across the three AUKUS countries is getting at buying and deploying military kit and solution sets that are gap fillers.

Last year I interviewed a senior U.S. Admiral who put this challenge very well:

Rehearsal of operations sheds light on our gaps. If you are rehearsing, you are writing mission orders down to the trigger puller, and the trigger puller will get these orders and go, I don't know what you want me to do. Where do you want me to be? Who am I supposed to check in with? What do you want me to kill when I get there? What are my left and right limits? Do I have target engagement authority?

This then allows a better process of writing effective mission orders, so that we're actually telling the joint force what we want them to do and who's got the lead at a specific operational point. By such an approach, we are learning. We're driving requirements from the people who are actually out there trying to execute the mission, as opposed to the war gamers who were sitting on the staff trying to figure out what the trigger pullers should do.

We focused in that conversation on the need for the ready forces to be able to be able to acquire short term capabilities to fill gaps. This will become crucial as we work the payloads on air and maritime autonomous systems to fill gaps.

Or put another way, a sense of urgency at this time in history requires an ability for greater acquisition authority to shift to the ready force.

And the interview I did with this Admiral also highlighted that the U.S. and its allies working together could shape a division of labor to fill gaps in addition to the acquisition of things, but this required training and operative force redesign among the forces working together.

This is what the Admiral emphasized:

We need to build a centralized planning organization in the Pacific, that has intelligence, future plans, future ops, current ops, ISR, logistics, all of those things included in the discussions. And it's made up of anybody who has desires to participate. This will allow us to understand what allied objectives are, and what their limits are, too, and helps us to identify the barriers such as foreign disclosure, so we can start to break down some of those barriers.

We can recognize weaknesses in our operational forces, the gaps as we have discussed, and we can focus on closing those gaps ourselves or by interchangeability with an ally. There are places where allies have capability that we do not have which can fill the bill.

Another point raised by the AIRMSHL was the need to fund and execute in a timely fashion the upgrades necessary for the core combat programs. This is especially important as air platforms have become software upgradeable, and ways need to be found not only to fund but to accelerate where possible software upgradeability.

For the ready force, there is an increasing need to be able to operate with significant software transient advantage, something that has been clearly demonstrated in the battles over electromagnetic spectrum in the Russian-NATO war in Ukraine.

A final point which I will highlight is training. The relationship between training and experimentation and inclusion of the outcomes of this effort INTO the ready force rather than on the conveyer belt of science projects for some future force is central to enabling a sense of urgency to lead actually to enhanced capabilities for the ready force.

Several years ago, I talked with Air Vice-Marshal (Retired) John Blackburn about this challenge.

This is how we discussed the challenge in this 2017 interview:

During my visit to Canberra, I had a chance to discuss with Air Vice-Marshal (Retired) John Blackburn how the training approach could be expanded to encompass and guide development.

"We know that we need to have an integrated force, because of the

complexity of the threat environment we will face in the future. The legacy approach is to buy bespoke pieces of equipment, and then use defined data links to connect them and to get as much integration as we can AFTER we have bought the separate pieces of equipment. This is after-market integration, and can take us only so far."

"This will not give us the level of capability that we need against the complex threat environment we will face. How do we design and build in integration? This is a real challenge, for no one has done so to date?"

Laird: "And the integration you are talking about is not just within the ADF but also with core allies, notably the United States forces. And we could emphasize that integration is necessary given the need to design a force that can go up against an adversary's military choke points, disrupt them, have the ability to understand the impact and continue on the attack.

"This requires an ability to put force packages up against a threat, prosecute, learn and continue to put the pressure on.

"Put bluntly, this is pushing situational awareness to the point of attack, combat learning within the operation at the critical nodes of attack and defence and rapidly reorganizing to keep up the speed and lethality of attack.

"To achieve such goals, clearly requires force package integration and strategic direction across the combat force.

"How best to move down this path?"

Blackburn: "We have to think more imaginatively when we design our force. A key way to do this is to move from a headquarters set requirements process by platform, to driving development by demonstration.

"How do you get the operators to drive the integration developmental piece?

"The operational experience of the Wedgetail crews with F-22 pilots has highlighted ways the two platforms might evolve to deliver significantly greater joint effect. But we need to build from their reworking of TTPs to shape development requirements so to speak. We need to develop to an operational outcome; not stay in the world of slow-motion requirements development platform by platform."

Laird: "My visit this year to the Naval Aviation Warfighting Development Center at Fallon highlighted the crucial need to link joint TTP development with training and hopefully beyond that to inform the joint integration piece.

"How best to do that from your point of view"?

Blackburn: "Defence is procuring a Live/Virtual/Constructive (LVC) training capability.

"But the approach is reported to be narrowly focused on training. We need to expand the aperture and include development and demonstration within the LVC world.

"We could use LVC to have the engineers and operators who are building the next generation of systems in a series of laboratories, participate in real-world exercises.

"Let's bring the developmental systems along, and plug it into the real-world exercise, but without interfering with it.

"With engagement by developers in a distributed laboratory model through LVC, we could be exploring and testing ideas for a project, during development. We would not have to wait until a capability has reached an 'initial' or 'full operating' capability level; we could learn a lot along the development by such an approach that involves the operators in the field.

"The target event would be a major classified exercise. We could be testing integration in the real-world exercise and concurrently in the labs that are developing the next generation of "integrated" systems.

"That, to my mind, is an integrated way of using LVC to help demonstrate and develop the integrated force. We could accelerate development coming into the operational force and eliminating the classic requirements setting approach.

"We need to set aside some aspects of the traditional acquisition approach in favor of an integrated development approach which would accelerate the realisation of integrated capabilities in the operational force."[1]

Let me close by examining more closely the AVM's comments on off-boarding.

When I worked for Secretary Wynne as a consultant beginning in

2004, I was focused primarily on the challenge of building a F-35 global enterprise. Wynne often made the point that a major problem for fighter pilots with fifth generation was that often they would not be delivering the kill shot, for their role was to identify targets for other shooters.

The off-boarding of weapons and sensors is at the heart of the kill web force, whether manned or inclusive of unmanned or autonomous systems.

This strategic shift to off-boarding is what allows one to consider massing force from different locations, from different platforms, from the joint force or the coalition force.

But how does the emergence of off-boarding affect the future of platform acquisition or the design of "next generation" core manned platforms?

In other words, adding new autonomous systems is not simply additive but goes to the core of weapons design and I would argue will change significantly future force design.

That is why I would argue that adding autonomous systems to a ready force which is built around a kill web con-ops of off-boarding is not simply a gap filler but a strategic shift in next generation platform design.

An F-35 force able to operate with autonomous systems lays the foundation for what follows, not what a force planner imagines will happen when such a capability is actually used.

5
FURTHER PERSPECTIVES OF AIR VICE MARSHAL GLEN BRAZ

I had chance to meet with Air Vice Marshal Braz and to further discuss with him the subject of the seminar in a meeting in his office.

I had the privilege of interviewing him when he was at Amberley Airbase commanding the Wing which hosted the new Growler squadron. And then again when he presented at the August 23, 2017 Williams Foundation seminar on the future of electronic warfare.[1]

Obviously, his experience with electronic warfare put him at the heart of evolving combat realities, namely the core significance of prevailing in the electro-magnetic spectrum.

During the discussion in his office, we went over the major elements of his presentation but discussed in more detail his point regarding the need to have a sense of urgency. It is crucial from a ready force perspective to drive change in how one organizationally shapes the force.

This is how he put it: *We need to enhance a sense of urgency in shaping the ready force.*

We've been very habitualized in managing long lead, large programs with long gestation and comprehensive oversight. We have done so through a slow approval process with rigorous analysis which deliver exquisite systems which allow us to shape a framework around which we can operate and with which we can operate,

But such an approach is lacking in terms of providing rapidly what the near-term enhanced force in being sees as immediate needs.

Such a system doesn't naturally deliver the enhancements that I would seek for the ready force in a timely manner. We need to adjust our focus on defence preparedness to encompass organizational change that can quickly deliver near-term solutions identified by our forces in being.

It was clear that as Air Commander Australia, Air Vice Marshal Braz has to be prepared for the ready force to support the missions directed by the Joint Force Commander. And one might put it that the adaptation of the force under duress is a key driver of change, and the challenge is organizationally how to build in the capacity to drive force design from such a perspective rather than doing so primarily from the standpoint of the core platforms to be acquired in the future.

And this approach underscores the importance of trusting the operational force to identify gaps to be met.

Air Vice Marshal Braz put it this way: *We talk increasingly about the workforce having a warfighter mindset. The ability and willingness for people to adapt is going to be fundamental to our war fighting effectiveness. I need every human that we have contributing to the effort, and I need them to be contributing in ways that they haven't even figured out apply to them yet.*

There's a trust equation inherent in mission rehearsal. In mission rehearsals, we find gaps, and we look for rapid ways, perhaps imperfect, to remedy the gaps.

And then we try to give technology that we think will solve known problems to our workforce, and we let them have the license to integrate it and operate with it in ways that we probably never imagined that they would.

But with their creativity and their adaptability they find ways to do so.

By then adopting the solutions they find into the ready force, we are incentivizing them to work in an innovative way that makes the ready force better in the near term.

I pointed out that when he was working in electronic warfare, it is an area which succeeds only by staying ahead of the game. In EW or more generally in tron warfare, it is crucial to ensure that the force has transient software advantage.

Air Vice Marshal Braz further emphasized: *One needs to be able to rebuild the force quickly, and that applies across the whole system of capabilities. We need to adapt quickly.*

And another element that I'm really interested in is our ability to generate tempo despite potentially disaggregating the force for survivability and for maneuver within that disaggregation.

But I need to be able to catch the adversary off guard and to punch multiple times in diverse ways. We need to be able to maximize our force and do things that will disrupt the enemy's thinking, and do the unexpected, which is going to be important for us a small force.

We need to be a smart force to which can provide technological advantage in the sense of being able to have ongoing and transient advantage.

Braz highlighted that collaboration among the Australian, American and British Air Warfare Centers provided an important way for the RAAF to explore concrete ways to get better value out of the ready force, and to cross-learn ways to gain combat advantages through enhanced training and identifying combat gaps that need to be and can be filled by technological or other fixes.

He argued: *We can find ways to seek asymmetry and apply new technology quickly. I thank such a capability is a key contributor to the deterrence equation.*

6

THE CHIEF OF THE AUSTRALIAN ARMY EXPLAINS HOW THEY ARE DEALING WITH THE SIMULTANEOUS CHALLENGES OF THE THREE R'S

In the September 26, 2024 Sir Richard Williams seminar, we focused on how the force in being can evolve effectively in the near to mid-term as investments are being made in future force capabilities, most notably the new SSNs.

This means that the services have to deal with three challenges simultaneously or what one might call the three R's: Re-structuring or redesign, readiness and resilience.

I had a chance to discuss how the Australian Army was dealing with these challenges in a meeting in the office of Lt. General Simon Stuart on September 29, 2024.

Lt. General Stuart underscored: *We must do all three simultaneously. That means that we have to change the way we are set up and the way we work.*

Previously, we did our force generation, force modernisation and readiness activities separately. They were three different parts of the force. We had a readiness model where we consumed readiness while we were deployed.

We cannot operate that way anymore, as we did in the so-called 'wars of choice'.

We start with the consideration that time is the key resource. It is insufficient to invest our time in three different activities and then converge on an operation.

We assign land forces to the Chief of Joint Operations (CJOPS), and I give him a level of assurance that these forces are ready and at a clearly defined readiness level. CJOPS assigns the mission and tasks for the particular operation, activity or investment. In our new way of working, I will also assign tasks in support of force generation and force modernization objectives.

When we deploy, whether onshore or offshore, bilaterally or multilaterally, we are going to make the best return on investment we possibly can.

For example, the forward deployed forces working in Indonesia as part of Exercise Super Garuda Shield worked with a partner and had assigned tasks from CJOPS and me. The tasks from me may be individual and/or collective training objectives or in support of force modernization. There might be an experiment. They may have new kit and I might task them to figure out what tactics, techniques and procedures we must adopt to employ and integrate this new kit on operations.

In other words, we are flipping the model from one where we consume readiness when we're deployed to one where we build readiness while we are deployed.

If we are doing that consistently across all operations, activities and investments, and we are doing that at scale, then we start to build readiness across the force.

We cannot afford a model where we have one part of the force at a high state of readiness and the rest of the force at low levels of readiness. It is very expensive and inefficient. We must be disciplined in understanding the difference between training levels and readiness levels.

I noted that the Army was a crucial force for working partnerships across the region, and frankly, I consider a major contribution of the ADF will be to enhance its operational capabilities out to the Solomon Islands. The Army can provide the kind of local knowledge and local partnerships crucial for the defence of Australia and for the region.

Lt General Stuart certainly agreed. *Some of the value of land forces is in presence and persistence, and those relationships you need in terms of placement, access and understanding the situation, the micro terrain, understanding the littorals, understanding the ports and the airports, understanding the language, the local culture. What does normal look like? And how do you detect what's different? How do you characterise threat?*

We then returned to a discussion of readiness built as well through a change in the training process.

Lt General Stuart underscored this approach as follows: *We have changed markedly since the so-called 'wars of choice'. Back then a battle group would go the Combat Training Center and be trained for a specific mission and theater. The trainers provided a full mission profile environment for the specific theater of operation.*

Today the battle group comes to the training center and is provided with a full mission profile for various operational environments and the Commander trains their unit or formation. That is how we are building readiness now.

We then shifted to the discussion of how to deal with the challenge of future force structure redesign.

Lt General Stuart emphasized the following: *In terms of force structure redesign, it is 18 months last week since the 2023 Defence Strategic Review. The Army has moved very quickly to execute on our mission and tasks, and our transformation.*

In that time, we have rewritten the land domain concept, the land operating concept (which translates the joint or the integrated force concept into the land force component), and translated the Chief of Joint Operations plans into force structure and readiness requirements for the Army.

We have rewritten a number of the subordinate concepts, for example, the special operations concept. We have re-organized the Army. We have changed its disposition, and we are getting after the reorganization of units at brigade and at battlegroup level, changed the way we do opera-

tional command and control and the physical footprint and disposition of our formation level headquarters.

We have created a dispersed nodal structure. We understand the bandwidth requirements, the data exchange requirements, the data standards and the architecture needed in order to operate in this manner. How does the Army contribute to and draw upon the combined kill web?

We have been exercising and experimenting over the last four years across northern Australia with the first brigade which is our lead unit for littoral operations. We've been doing that across the North of Australia and projecting into the Northwest. We've been doing that with our teammates in the Marine Rotational Force-Darwin, and U.S. Army Pacific and specifically with their composite watercraft company.

We have leased civilian stern landing vessels to practice and to experiment and figure out how we're going to incorporate new weapons, new watercraft, new digital systems in order to meet our operational mission.

It is experimentation with a focus on 'learn by doing' and builds readiness in the process.

We are redesigning the Army in a very practical way. And the way that our soldiers have embraced innovation from the ground up to solve operational problems is just phenomenal.

It's a work in progress, but it's moving quickly, and we are working with every partner, whether industry, allies or the other services to get after these problems. In that way, the redesign turns upside down the capability development and delivery process.

We used to start with the major system, let's say watercraft. We then built some facilities, we trained some people, we did some Operational Test and Evaluation, and then we fielded the system. That process would traditionally take about a decade for a major system.

One of the requirements of the 2023 Defence Strategic Review is to change the way that the Government, Defence and the other agencies do acquisition. While we wait for those changes to occur, we're doing what we can with what we have and taking that approach already.

For example, the very last thing to be delivered for the littoral maneuver capability will be the watercraft. We'll have the doctrine, the concepts, the tactics, techniques and procedures already adopted.

We will already have adopted different structures, different ways of working, and different equipment sets to support how our formations fight.

The third piece, which you asked about, was resilience. We are not going to fight alone nor are we going to sustain ourselves on our own. We are working with small and medium enterprises, Australian enterprises notably, as well as the large primes we are associated with, to build the magazine depth and effectors we need for today and tomorrow's fight, particularly when it comes to long-range strike.

The other aspect to resilience is what I call 'adaptive reuse'. In other words, what do we have that can be reused in different ways – perhaps with a technical inject? Because you go with the kit you've got in a 'fight tonight' situation.

How can we use our extant kit in different ways through the application of technology or by integrating it into a human-machine team?

We are building an ecosystem that fosters innovation from the ground up, adding resources to it, and we're getting some great results.

We have completely changed the mission of our 1st Armored Regiment which was previously a tank regiment, but it is now the lead trace for applied modernisation in our Army. It is the center of a network of industry, military and academia focused on solving today's problems by putting new kit in the hands of our soldiers and enabling them to figure out how they are going to best use it operationally.

7
THE PERSPECTIVE OF AUSTRALIA'S AUKUS PARTNERS ON SHAPING THE WAY AHEAD FOR AIRPOWER

General Kevin Schneider, Commander of U.S. Pacific Air Forces and AVM Mark Flewin, Air Officer Commanding 1 Group Royal Air Force, both spoke at the September 26, 2024 Sir Richard Williams Foundation Seminar.

The first presentation was by a video recording and the second was an in person presentation after the long flight from the UK. Together, the two provided insights into USAF and RAF thinking about the way ahead with allied airpower.

General Schneider underscored the close working relationship between the U.S. and Australia over the years and provided several examples of recent collaborative activity. He focused on the recent Pitch Black exercise which was especially notable because of the expansion of partners in the Pacific region who participated.

Exercises like Pitch Black are not only increasing our interoperability, but they are helping our allied and partner nations rapidly to grow their capabilities. This in turn, helps secure their nations and provide stability to the region.

Your leadership in that exercise was evident, and the work accomplished there reached audiences around the world. You brought in so many firsts, the deployment of the Philippine Air Wing, our partners in Papua, New Guinea, and you worked tirelessly to bring in critical NATO allies from France, Spain, Germany, Italy and the UK.

NATO recognizes the importance of the Indo-Pacific and understands the impact it has on Euro Atlantic security. They also know where to find world class training with allies like you as well.

He then when on to highlight Australian cooperation in Filipino training as well as a key contribution to the common defence in the region.

I found quite interesting his spending time discussing the E-7 as well. When I first came to Australia in 2014 and then subsequently visiting the RAAF Wedgetail squadron, it was clear that the ADF had something special in the Wedgetail. But the United States and the United Kingdom even though looking for AWACS replacements were slow to embrace the Wedgetail option. But now they have.

He noted: *The RAAF's significant focus and role in air domain awareness stems from your nation's early investment in the E-7 Wedgetail. It is a critical asset in our most advanced high-end training. The U.S. Air Force Weapons School hosted the RAAF's No. 2 Squadron for the first time as part of the weapons school integration phase in May and as part of the modernization of our own fleet.*

We are excited to see the expansion of the exchange program, because now No. 2 Squadron is a multinational, integrated unit whose regular participation in global exercises with joint partners is a must, because we are developing and testing E-7 tactics with their air superiority and support for maritime strike forces. We wouldn't be able to accomplish any of this without the genius of our collective airmen who are doing wonders at the tactical edge.

The General underscored: *Australia adds immense value at the cutting edge of our most advanced tactics, techniques and procedures.*

Quite obviously, the USAF needs to operate very differently in the evolving contested combat environment. And in this effort, it is working hand in glove with the RAAF.

This is how PACAF put it: *We must reorganize ourselves to tackle the high-end fights in the future, where we must be lethal while surviving in an anti-access area denial environment. We are learning to operate from austere locations, testing critical capabilities like our bomber Task Force and stressing our agile combat employment concept through a series of complex exercises at scope and at scale.*

Through tremendous support from you, we've increased the rotational presence of U.S. capabilities in Australia across all domains, ensuring our forces work as interchangeable teams who are efficient during peacetime and lethal and survivable during wartime...

We want to find more ways to operate from different locations around the region to drive solutions to logistical challenges and to conduct rehearsals like hot pit refueling events and integrated flying operations to make our footprints even more lean and agile.

The bomber task forces, and our strategic aircraft play a critical role in our collective ability to support counter-maritime missions, something that we must do because the rise of competitors in both the Indo, Pacific and European theaters has brought anti-ship capability back to the forefront of the ASW mission. The bomber fleet is finding innovative ways to integrate modern weapons capabilities to increase survivability in an anti-access area denial environment and to support the joint fight.

He then discussed Agile Combat Employment and immediate ways ahead on training for this capability. *Next summer, we'll partner with like-minded nations to host a large scale exercise to test agile combat employment at speed and scale in the Pacific that will coincide with the Talisman Sabre exercise, and as we anticipate the exercise will include fifth generation fighters, ISR, C-2, airlift and air fueling, and all the enablers to test our ability to deploy from the continental United States into theater to regional hubs in the first and second island chains.*

We will disperse, aggregate, disaggregate, and recover aircraft. It is a highly complex logistical challenge in terms of access, spacing and overflight, maritime domain awareness and maritime strike capabilities, as well as generating and sustaining the force, making it even more challenging, we are adapting this new operational scheme of maneuver under significant fiscal constraints, a challenge that we all face as exciting as all these things are, I will never say that what we are doing is fast enough, that we have integrated enough, that we have prepared enough, or that we are ready enough.

AVM Mark Flewin then provided an RAF perspective on the way ahead.

AVM Flewin presenting at the September 26, 2024 Sir Richard Williams Foundation Seminar.

He conveyed the sense of urgency as the West faces increasing threats and challenges. He looked back at World War II to remind the audience of the cost of the failure of deterrence.

But his presentation underscored the need for the West to get it right in terms of deterrence and although there is clear progress progress in the West's capabilities, the tenor of his remarks that as an enterprise, we need to get better in order to ensure that deterrence prevails.

He underscored what he saw as five critical challenges that needed to be met on an urgent basis.

The first is the need to generate greater combat mass.

He noted: *We need to have the capacity to scale. It doesn't need to be exquisite in terms of combat systems, though it can be. It needs to be on the right side of the cost curve. It can be cheap. We've seen from Ukraine that there's a heavy mix of exquisite and non-exquisite capability that is ultimately delivering effects. But we need to absolutely partner with industry to be able to do that.*

The second is enhancing our ability to fight tonight. The majority of the equipment we will have in 10 years we already have so we need to engage in continuous force improvement and training to ensure our force in readiness is at the level needed for deterrence

He underscored: *We need to work together to optimize and get the most out of our platforms. We're working that in the Royal Air Force through a program called Optimize. It's seen significant benefits already. We've seen the 20% improvement in availability on some platforms.*

With Typhoon, for example, we've managed to remove 750,000 maintenance hours from that platform based on some data analysis and a risk aware approach which means we get more availability, and our mechanics are available for other tasks.

And it's really important that we continue to spiral develop these platforms. They are going to be the baseline of our capability up to 2040 and they need to be ready to deliver.

The third is to ensure that we can adapt rapidly to technological change and to be able to incorporate relevant combat innovations being unveiled in the regional wars we see in front of us.

He put it this way: *My next point is on the criticality of embracing technological change. We talked about the electromagnetic environment today. What we witnessed in eastern Ukraine is that it's an absolutely denied electromagnetic environment, we need to get around how we work in that environment, how we evolve in that environment and how we can bring operational advantage in that environment. We also need to not be afraid to fail and test and fail fast. It's something we've worked on a lot in the UK. Naturally, we've moved away from it because we've been risk averse.*

The fourth is the challenge of overcoming risk aversion and become more agile and capable of rapid innovation in our tactics and warfighting skills.

AVM Flewin emphasized: *We need to continue to change our*

mindset and make sure that we're ready for the fight tomorrow. You might throw back at me that our processes aren't efficient enough.

We procure very slowly, our commercial process isn't proficient enough. We've learned a lot about that through Ukraine.

We've adapted our process with industry.

And I'd argue now that we are, we are getting after it, but there's more to do. And we clearly need to transition from risk aversion to risk aware and objective focus.

8

KEY WARFIGHTING CAPABILITIES FOR THE NEW STRATEGIC ENVIRONMENT

The strategic environment has clearly changed and the level of real world conflict has escalated, whether in terms of major hot wars – in the Middle East and in Europe – or in terms of active gray zone conflicts.

The ready force needs to deal with this real world rather than than just preparing for a future war. The Defence Strategic Review (DSR) has outlined the future force for the future war; the current ADF needs to deal with the world we have.

The prospect of armed conflict in the first island chain, already underway in the gray zone between China and the Philippines, and the prospect of PRC actions against a sovereign free nation in Taiwan has been constantly threatened by the current leader of China.

How do you respond to the here and now but do so in a way that puts you on the trajectory which the DSR has mandated?

Some answers to this question were provided by three of the speakers at the seminar.

Their answers were in the form of identifying capabilities crucial to the ADF now which needed to be underscored and enhanced in the years to come.

Lt. General Susan Coyle presenting at the September 26, 2024 Sir Richard Williams Foundation seminar.

The first speaker was Lt General Susan Coyle, Chief of Joint Capabilities. The Joint Capabilities Group is defined by the Australian government as follows:

Joint Capabilities Group (JCG) is headed by the Chief of Joint Capabilities (CJC), who is responsible to the Chief of Defence Force for the provision of Joint Health, Logistics, Education and Training, Information Warfare and Joint Military Police. CJC will also manage agreed Joint projects, and their sustainment, to support joint capability requirements.[1]

Concomitant with her broad remit, Coyle discussed a range of efforts within the JCG. But the most important for this book was what she identified as the "high ground."

Cyber power is a vital element of national military power. We need to coordinate electronic warfare and cyber information operations in order to gain asymmetric advantage and paralyze our adversary's decision making. We must be able to continue to defend and exploit capabilities within the electromagnetic spectrum.

And I've heard this referred to recently as the next high ground. Doing so will slow adversaries kill chains and increase confusion and degrade their precision. The layering of electronic warfare, cyber and kinetic attacks is similarly vital for us to assuring our own strike capability.

We must protect radios and microwaves that are used for communications and radars, as well as infrared spectrum for our weapons guidance, jamming aircraft, blinding air defence, radars, suppressing military missile systems. All of this is absolutely real and at the forefront of our mind, and so we're focused on delivering capability that will protect our ability to operate across the electronic magnetic spectrum.

Lt General Coyle put it bluntly: *If we lose the war in the electromagnetic spectrum, we lose the war across all domains.*

Hence enabling the force in being to fight, survive and prevail in the current context is crucial and in so doing carves a path to shaping future options as well.

The second speaker weighed in on Coyle's comments. The speaker was Phil Winzenberg, the Deputy Director General-Signals Intelligence and Effects from the Australian Signals Directorate. He reinforced Coyle's emphasis on the role of effective operations in the electromagnetic spectrum and signals intelligence as enablers for an integrated force.

In his presentation, he underscored that the integrated force was more than the ADF and its operations.

This what he argued about the ASD and its current relationship with the ADF in creating a more integrated force: *That brings us to the present, where the demands from the ADF have increased ever further. Instead of just the land domain, which was the focus of previous conflicts, we will support the ADF in all domains. We won't just provide intelligence on adversary capabilities, but also timely indicators and warnings of adversary intent towards the integrated force and also our nation's critical infrastructure.*

We won't just provide situational awareness of the disposition of adversary platforms and forces, but we'll do that also to establish and hold custody of them so that they can be use effectively by CJOC, at his or her discretion, and on a scale and with a complexity never encountered before by ASD or indeed the ADF.

Phil Winzenberg presenting at the September 26, 2024 Sir Richard Williams Foundation seminar.

He then highlighted REDSPICE.

This programmatic effort is described by the ASD as follows:

REDSPICE is the most significant single investment in the Australian Signals Directorate's 75 years. It responds to the deteriorating strategic circumstances in our region, characterised by rapid military expansion, growing coercive behaviour and increased cyber attacks.

Through REDSPICE, ASD will deliver forward-looking capabilities essential to maintaining Australia's strategic advantage and capability edge over the coming decade and beyond.

The REDSPICE Blueprint (PDF) offers some further insights into what REDSPICE will deliver, and a vision of what ASD will look like in the future.[2]

Through REDSPICE, we will expand the range and sophistication of our intelligence, offensive and defensive cyber capabilities, and build on our already-strong enabling foundations.

- *3x current offensive cyber capability*
- *2x persistent cyber-hunt activities*

- *Advanced AI, machine learning and cloud technology*
- *4x global footprint*
- *1900 new analyst, technologist, corporate and enabling roles across Australia and the world*
- *40% of staff located outside Canberra.*[3]

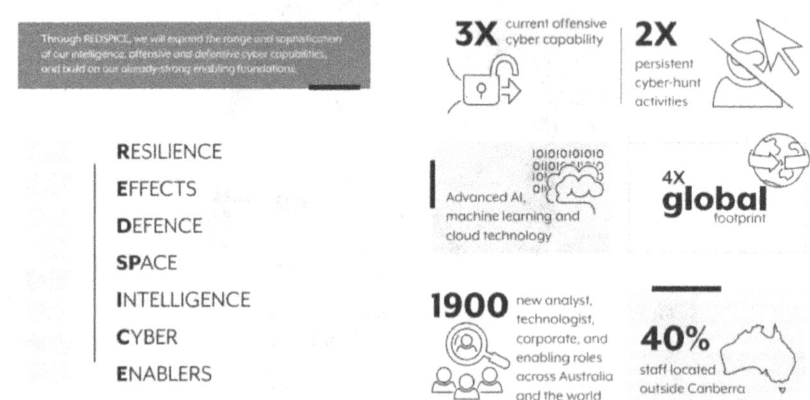

Winzenberg went on to underscore the following: *I can assure everyone here that delivering REDSPICE is an ASD top priority, and we're doing all that we can to hold up our end of the bargain in supporting the future integrated force.*

REDSPICE is supporting current and future operations from the strategic to the tactical. There is no defensive cyber activity that the integrated force plans or executes that doesn't draw its threat intelligence or employ tools developed by REDSPICE.

There is no information and cyber effects that the integrated force plans or executes that doesn't happen without REDSPICE targeting intelligence and tools, and there's no technical intelligence on weapon systems and other capabilities that allow us to develop countermeasures and build electronic attack algorithms that isn't enhanced by REDSPICE.

He then underscored that the ADF-ASD partnership was enhanced by the Five Eyes alliance.

I'll finish by reflecting on the power that comes from ASDs membership of the five eyes. The Five Eyes alliance is the greatest intelligence partnership the world has ever known...

The trust and depth of this partnership is the key differentiator between us and others in the region. The breadth and depth of what we do together is truly staggering.

If we set about trying to achieve that today from the standing start, it would be inconceivable that we would get to where we are now. We have good, good friends everywhere who willingly work with us. They work with us to build our capabilities. They work with us to secure our interests. They work with us to share intelligence so that we can understand the challenges of the world together and in similar ways.

He then concluded with this statement: *So to close, I'm going to leave you with three fun facts.*

Number one, ASD is at its core a support to military organizations. We're now driven to deliver REDSPICE to best support government and defence in our complex strategic environment.

Number two, partnerships are key to the integrated force, and the ADF has no better partner than ASD and our five eyes buddies as we all rise to the shared challenge.

And number three, last but not least, when it comes to actionable intelligence at the speed of relevance in a modern battle space, you only have two options, SIGINT or everything else, much, much, much too late.

The third speaker was the Chief of the RAAF, AIRMSHL Stephen Chappell. He focused on the importance of taking advantage of the geographical situation in which Australia finds itself.

In effect, by shaping the RAAF and the integrated force as a maneuver force, the ADF is able to use geographical depth to defend itself and at the same time able to deploy from various locations on its own continent to gain strategic and tactical advantage.

AIRMSHL Stephen Chappell presenting at the September 26, 2024 Sir Richard Williams Foundation seminar.

This is how he concluded his presentation:

We can create the tempo, confuse and complicate adversary targeting and project depth well beyond their shores by working together as a joint and integrated force working closely with interagency, industry partners, allies and partners. We can position and posture to deter and be prepared to respond. And I'd ask you to as we roll out the concept aspect, embrace it.

What he was articulating was deterrence by maneuver in terms of geographical and operational (meaning an ability to leverage space and extended range ISR and C2 systems) depth.

He highlighted this approach as follows: *Going back to how to generate our depth. We've experimented using space and cyber in generating effects to gain advantage. We've integrated with the U.S. and Australian fleets and land forces across the primary area of military interest for us to execute multi domain strikes at range in defence of our territory and coalition forces.*

Our war games show us how in the Indo Pacific we can effectively integrate capabilities and forces across domains and nationalities, operationally and strategically. They also show us that there is much more to be done, and therefore our services will have to continue to evolve to meet the requirements of multi-domain operations.

He provided further details in a earlier presentation as well. In the 27 September 2023 Sir Richard Williams Foundation seminar, he made the following comment:

AVM Stephen Chappell, Head of Military Strategic Commitments, focused on the symbiotic relationship between offense and defence with regard to multi-domain strike.

"Integrated air and missile defence and multi-domain strike are two sides of the same coin. By having an ability to protect our strike enterprises we enhance their credibility to strike back which ensures as well our enhanced deterrent capability."

He underscored: "Passive defence is just as important not only for defending critical assets, but preserving our multi-domain strike capabilities in order to execute those left jabs and those right hooks necessary. The next layer of defence we're thinking about is counter force.

"This in effect is multi-domain strike, the ability to reach out and defeat a threat to our homeland or to our forces. Deterrence by denial includes that defensive protection of the chin as we deliver effective left jabs and right hooks."[4]

9

DEALING WITH A RAPIDLY CHANGING OPERATIONAL ENVIRONMENT

It is fine to have long-range force structure planning, but what happens when the operational environment is rapidly changing for your operational force?

How to adapt the ready force effectively and adeptly in a timely manner?

And what consequences does that have for one's long-range force structure design?

The presentation by Jennifer Parker, Expert Associate National Security College of the Australian National University, focused on a key challenge which raises such questions. Her presentation was entitled: "The Contested Maritime Domain: Challenges for an Integrated Force?"

What she focused on was the changing nature of littoral maritime operations, the emergence of new technologies and concepts of operations by various actors notably in the Black and Red Seas, and how those shifts in approach affected maritime operations.

The bottom line of her analysis was that the new technologies and approaches had a clear impact on capital ship naval vessels, and with relevant defence measures, technologies and relevant training, capital ships could still operate effectively in the littorals.

But the point can be put bluntly: you need to adapt your ready force to deal with new technologies, new con-ops and technologies.

And a point outside of her presentation was inherent within it: what is the future of capital ships integrating maritime autonomous systems? For defence? For offence?

Or as I would put it, it is not a question of crewed versus uncrewed vessels. It is about how crewed vessels could leverage off-board assets like maritime autonomous systems or air systems for the projection of effect or defence in depth.

Jennifer Parker speaking at the September 26, 2024 Sir Richard Williams Foundation seminar.

Her first case study was of the Black Sea and contested littoral operations there.

She argued that: *The range of the littoral is increasing. Ukraine has effectively used uncrewed surface vessels, cruise missiles and UAVs to target ships at greater ranges.*

Now we don't know the exact ranges of some of the uncrewed surface vessels that Ukraine has operated, but certainly they managed to hit the

Kursk bridge at about 300 nautical miles from Ukrainian controlled territory.

That is a dramatic change in terms of the range of the littoral. Ukraine has managed to destroy about 30% of Russia's Black Sea fleet, and certainly pushed them back from Ukraine and territorial seas and from Crimea, and that's no mean feat when you look at the USV engagements,

But what must considered is how the Russian ships have defended themselves. They have no countermeasures whatsoever. They're barely maneuvering or defending themselves...As offensive capability evolves, we need to be working hard on what the defensive capability is and integrating it into our platforms.

She noted as well that USVs can certainly attack ports and port infrastructure. This means that there need to be countermeasures for this new threat as well.

The key takeaways from her analysis were the following:

- Range of the littoral is increasing:
- Sea denial strategy in enclosed seas
- Ship preparedness / posture is key
- Balance between development of offensive / defensive balance (capability / counter capability)
- Maritime trade does not stop during conflict
- Importance of port infrastructure protection.

Her second case study was of the Red Sea and the approach of the Houthis to disruption in the littorals.

She argued that: *The Houthis have been successful in changing the direction of maritime trade. There have been over 100 attacks now on merchant shipping, and 30 of these attacks have managed to sink a couple of ships. This shows the vulnerability of choke points using the kind of systems and technology available to the Houthis. They have attacked but not damaged surface combatants.*

She underscored that prepared surface vessels have successfully defended themselves but two problems have been underscored for the ready force. First, the fleet needs to find ways to be rearmed with missiles while at sea. Second, the fleet needs to find much cheaper ways to defeat the unmanned strike force directed against the fleet.

She argued that it was necessary for the ready force to be "stressed tested" by engaging in such deployments to evolve its combat edge.

The key takeaways from her analysis were the following:

- Vulnerability of chokepoints
- Continuing relevance of surface combatants
- Magazine depth / Replenishment at sea
- Integration of counter-drone capabilities
- Importance of stress testing capabilities
- Continuing relevance of convoy operations
- Strategic depth in maritime fleet
- Defence / maritime industry coordination
- Maritime trade doesn't stop / it evolves.

Her third case study was of the PRC actions against the Philippines. The key takeaways from her analysis were the following:

- Blurring of civil maritime security threats
- Criticality of maritime domain awareness
- Effective presence operations require quantity of forces
- Integration of information operations into wider campaign strategy.

And finally, she addressed the active threats to sea laid cables which are critical to the information flows globally. Here she asked the poignant question: "Whose responsibility for such defence is this in Australia? And what are we doing about it?"

This challenge is a key one, which parts of NATO are finally addressing in Europe. For example, in the recent NORDIC WARDEN Exercise, the UK and Northern European nations exercised their forces to shape a con-ops to deal with this, although the exercise indicated important technology and force structure gaps to deal with the challenge.[1]

With regard to operations in the Red Sea, she noted: *A number of the European navies who have gone through their workups and gone through their test and evaluation have sent ships to the Red Sea and learned very quickly that their combat systems and their missile systems were not up for the fight and had to withdraw them.*

That is something that we want to learn before we are getting multiple missiles shot at us in the event of a more significant conflict.

Parker underscored the really crucial point that when it comes to naval operations, the military and the civilian aspects are intertwined. Australia depends on maritime trade, which will need to continue in times of conflict, and to do so, the military and civilian parts of the equation need to be clearly working together.

She noted: *We've learned that defence and maritime industry need to coordinate now. That's something that we consistently relearn, and that has been a key point of the defence of merchant shipping in the Red Sea, and something that Australia needs to think about as we try and grow our maritime industry with strategic fleet...*

Maritime trade does not stop in the event of conflict, so the view that we don't need to worry about it or that we just need to worry about protecting Australia has not borne itself out in our previous world wars and it is not bearing itself out in the Black Sea or the Red Sea.

And what is happening in the west Philippine Sea is a clear blurring of civil maritime security threats. This is something that we need to pay attention to.

We currently have a civil maritime strategy. We have a military maritime strategy, and the two don't connect. It's not too far to think that an adversary could try to overwhelm Australia's maritime domain through using what we would continue consider civil maritime threats.

10

PROTECTING THE NATION: ITS MORE THAN THE ADF'S ROLE

We focused in the seminar on ways for the force in being to be augmented in the short to midterm.

But for the effective defence of Australia as a nation, one needs to expand the notion from a force in being to a force in being embedded in the national enterprise which participates beyond the remit of the ADF to defend the nation.

Two presentations at the seminar provided insights in how to think about this approach. The first was by Chris McInnes, Executive Director, Air Power Institute, and the second was by Dr. Malcolm Davis from the Australian Strategic Policy Institute.

McInnes highlighted the importance of an Australian-wide aviation enterprise incorporating the civil sector as a key part of the overall defence of the nation. Davis focused on the space dimension as a key part of Australia having an effective capability to participate in the broader information society domestically and internationally.

In this chapter, I will address the McInnes presentation and in the next chapter the Davis presentation.

Chris McInnes speaking at the Sir Richard Williams seminar on September 26, 2024.

McInnes entitled his address: "Building a national air power enterprise."

This is what he had to say on this subject:

Australia needs a concept and vision for a national air power enterprise. This is National Defence – the foundational principle of the NDS – in the air and is key to accelerating and enhancing the ADF's effectiveness.

This enterprise includes the civil and military aviation sectors as well as the aerospace industry as complementary and interdependent components.

We need a vision of where the nation needs this enterprise to go, with national priorities and policies to reshape Australian air power.

These are the first steps in Australian air power becoming more than the sum of its parts.

We need it to be more than the sum of its parts this because Australia is a nation "uniquely reliant on aviation" as the Government's Aviation White Paper declared in August.

We need it to happen now because the air superiority underpinning that reliance could be disrupted at any moment and responding to that challenge will be a national effort.

We are four years along from when we last had ten years warning time.

Historical analogies are imperfect but as Prime Minister Morrison invoked the 1930s when discarding Australia's 'ten-year rule' in 2020, so shall I.

Britain dropped its ten-year rule in March 1932 – after Japan's invasion of Northeast China but before Hitler came to power in Germany. By the end of 1936, four years after Britain's ten-year warning time elapsed – that is, where we are now - it had:

- *Begun coordinating production of airframes and aero engines through the Air Ministry;*
- *Devised the shadow factory scheme, which built more production capacity 'in the shadow' of related industries;*
- *Flown the first prototype Spitfire in March 1936 and ordered the first 310 production models by year's end;*
- *Massively expanded flying training, including forming the RAF Volunteer Reserve and using civilian flying schools for elementary training.*

Many setbacks lay ahead, but the foundations of Britain's air power enterprise were in place.

This included the "foundational principle" that air power was a national – indeed an imperial – enterprise for Britain.

This enterprise approach is why the multi-national RAF never came close to losing control of the skies over Britain and why Western air forces eventually dominated the skies in all theatres, despite their opponents' head start.

I would ask you to ponder the status of Australia's national air power enterprise four years since our own ten-year rule lapsed.

While you think on that, I will to talk about why air power and aviation matters specifically to Australia.

When it comes to aviation and air power, I suggest many Australians – including large swathes of the Defence organisation – are bit like the two young fish in David Foster Wallace's anecdote. After a passing older fish says, "morning boys, how's the water?" the two young fish swim on for a bit before one turns to the other and asks "what the hell is water ?"

This is an aviation nation – but we take it for granted.

The Aviation White Paper described Australia as "uniquely reliant on aviation" because it said we are, "a vast island nation with a dispersed population, far from our key trading partners and visitor markets, air transport provides Australians with critical links to each other and the world."

Australia has about 0.3% of the world's human population but responsibility for 5% of its surface area.

Air power turns days into hours and gives Australia and its residents choices and abilities that would not otherwise exist.

This is why Australia was an early adopter of aviation services – such as Qantas and the Royal Flying Doctor Service – and remains dependent on them.

The same is true militarily. The ADF reduces its need for surface forces by relying on air- delivered firepower, communication, transport, and observation.

Both civil and military aviation depend on air superiority to use airspace free from prohibitive interference. Air superiority is the primary reason governments formed separate air forces– to ensure a nation could harness its air power to first secure its skies.

A key implication of the ten-year rule lapsing is that Australia's comfortable assumption of air superiority is no longer valid, even over the homeland. And it will not take a major conflict to challenge air superiority because it is generally a subjective belief. The perception of risk from any source can be enough to disrupt aviation, often severely.

Volcanoes, military operations or exercises, security incidents, and aerial intrusions have all disturbed the free use of airspace in recent years. Consequently, potentially hostile actors, such as the Chinese Communist Party need not actually attack or threaten Australia's airspace or aviation to unsettle nerves. Consider the experiences of Taiwan and Japan in recent years.

While not without challenges, the Peoples Liberation Army can project power into Australia's periphery in a straightforward and perfectly legal manner. Its about 2,500 km from PLA bases in the West Philippines Sea to entering the Indian Ocean via the Makassar and Lombok Straits. That is about 3.5 hours flying time for H-6 bombers. Another two hours has them approaching Australia's coast anywhere from Exmouth to Darwin . Its less than a week's sail for a PLA naval task group – which could include aircraft carriers approaching the size of U.S. super-carriers in the next few years. Notably, last week saw the first open-source evidence the PLA had three aircraft carriers at sea simultaneously. Six will be in service by 2035, including four big ones.

The world has seen over the last few decades how sensitive Australia's politics can be to border security.

Consider the outcry over images of PLA aircraft with Darwin in the background. Imagine they are visibly armed. Now imagine there are no Australian aircraft with them. In his classic 1966 book The Tyranny of Distance, Geoffrey Blainey observed "the ease with which foreign bombers... could fly to Australia was probably the sharpest mental change" for Australians experience of aviation in the first half of the twentieth century.

The same seems likely to hold true this century.

Now I have no doubt the RAAF can monitor and intercept PLA aircraft approaching Australia's airspace. But covering multiple areas or sustaining alert for weeks or months if not years must surely be a different matter.

A peacetime challenge to Australia's air superiority would have national implications and require a national response. An actual conflict would only compound this.

The elements of this response exist but there are few public signs of a coherent concept of an Australian national air power enterprise to add necessary depth.

The NDS, released in April, called for 'a whole of government and whole of nation approach to Australia's defence' and that we needed National Defence as "a concept that harnesses all arms of Australia's national power." Unfortunately, nobody told those responsible for the Aviation White Paper, published in August. Instead, we have departmental rather than national policies that pass each other by like aircraft in cloud. Each treats the civil, military, and industrial components of Australia's air power as separate entities rather than part of a coherent whole.

The AWP – produced by the Department of Infrastructure, Transport, Regional Development, Communications, and the Arts – focused tightly on civil and general aviation.

The Department of Defence is not listed as having made a submission to the AWP, though it may have done so confidentially.[1] *The NDS dealt with defence issues, of which air power is but one.*

Policy responsibility for the aerospace industry, meanwhile, is truly fragmented – resting across Defence – which published the Defence Industry Development Strategy in February – and the Department of Industry, Science and Resources.[2]

Thinking of military power as a national enterprise is not unique or new, it just is not clear in Australian air power policy.

The ADF peak air power doctrine defines it as "the total strength of a nation's capability to conduct and influence activities in, through and from the air to achieve its objectives." This explicit characterisation of air power as national capability is a promising start, but the doctrine otherwise focuses solely military aviation and its applications.

Navalists have consistently portrayed Australian sea power as a national enterprise for years, leading to government investment in Australia's merchant fleet and naval spending consuming more of the Defence budget than air, land, and cyber forces combined.[3]

Britain's 2018 combat air strategy explicitly placed a national values framework including whole-of-nation security and prosperity at the heart of its decision making.[4]

New Zealand's national aerospace strategy released this week explicitly links a vision of doubling the size of aerospace industry by 2030 to the country's security and foreign policy.

Australia's fragmented approach to air power is is making the enterprise less than it could be, just when shared challenges and deteriorating security demand the opposite.

There are shared challenges across the enterprise for which national solutions may offer opportunities. At the highest level, there are over 16,000 aircraft on Australia's civil register performing a wide variety of roles, including surveillance, response, and airlift.

What contribution could these make to Australia's defence? What mechanisms are needed to make that happen?

To zoom in on a specific example, there are 28 A330 type aircraft listed on Australia's civil register. This is the platform upon which the RAAF's seven overworked KC- 30 tankers are based.

How quickly and by what means could civil A330s be modified in Australia to boost refuelling capacity?

What about A330 crews? Can they be inducted into military service to add depth or free up permanent crews for other requirements? Do they have to be in uniform to do so?

Zooming out again, there are more than 50,000 Australians working in civil aviation – that is a pool of aviation-savvy Australians almost triple the size of the RAAF. What role can they play in the defence of the country?

Workforce shortages are among the most serious challenges facing every part of the air power enterprise. But the components are competing for a diminishing resource.

The AWP recognises the ADF is a major source of skilled aviation labour but then describes it as a competitor that will "exacerbate future skilled aviation workforce challenges in Australian civil aviation."

Treating the aviation skills base as a national asset could create opportunities to address civil shortages in the short-term while building a trained reserve for times of crisis, and the mechanisms for their employment.

Such an approach could look like the Volunteer Reserve approach Britain employed in the 1930s. This low readiness reserve supported part-time aircrew training through civil aviation schools while also providing a means to induct trained personnel into service should the need arise. The use of military training arrangements to grow the civil workforce while building a military reserve illustrates the interdependency of the air power enterprise.

Infrastructure challenges are also common. The NDS says the Govern-

ment will spend up to $6.6 billion on the ADF's airbases in the next decade and acknowledged the need to ensure "civil society and civil infrastructure can support ADF requirements." This may be implemented through a Northern Air Base Network, alluded to at a conference in May, that will include "established military bases as well as other places that can support expeditionary air power."[5]

These NDS objectives are relevant to the AWP initiatives to spend an extra $90 million on the Remote Aviation Access Program and Regional Airports Program over the next three years.

But on this, along with every other aspect of civil aviation's significance to the defence of the country, the AWP is silent.

What is missing altogether is guidance on how the Australia's aerospace industry – development, manufacturing, and repair – can reshape Australia's air power to create new potential in the national interest.

Instead, the aerospace industry is viewed primarily as a supporting function – and policy and priorities, often conflicting, are spread across multiple sources.

The Defence Industrial Development Strategy alone spreads direction for the aerospace industry across at least three of seven priority areas.

According to a 2019 report commissioned by the Federal Government, Australia's aerospace industry included almost 1,000 companies, employed 20,000 Australians and boasted world-class research quality, uninhabited systems, and advanced manufacturing.

This is far more than a maintenance and repair industry.

Critically, the report also found the aerospace industry is independently commercially viable through diverse customers, including exports. This is not a government monopsony.

The report found the aerospace industry added almost $3 billion annually in gross value to Australia's economy – more value add than shipbuilding and rail rolling stock – which are both supported by national approaches.[6]

In the five years since then, and despite Covid, Australian companies have gone onto design and build a growing variety of advanced aviation components and autonomous systems domestically.

These include the fighter-sized Ghost Bat combat aircraft, designed, and built by over 200 Australian companies and the electric Vertiia – a vertical take-off and landing aircraft whose planned hydrogen-fuelled version will carry 500 kg payloads over 1,000 km.

In Australia's geography, size matters.

But if there has been Government support for these efforts, it has been platform – or project-based, rather than efforts to build a coherent Australian autonomous air power ecosystem.

In a speech in May, the Chief of Air Force said the Air Force was exploring autonomous air systems to build a "national ecosystem that can rapidly scale production of uncrewed systems." That is promising but surely we can be doing more than exploring. Australian industry has a demonstrated competence and comparative advantage in a field that is advancing rapidly around the globe.

Moreover, it is an area of particular value to Australia because autonomous systems could liberate Australian air power from the constraints of a small population, just as aircraft overcame the tyranny of distance a century ago.

This is what I mean by the aerospace industry reshaping air power to create new potential.

But we first need a coherent concept of Australia's air power enterprise and a vision of where the nation needs it to go. Australian aviation operators and aerospace companies can then compete or collaborate across the enterprise to meet those needs. Governments should guide and support but avoid trying to control a a market-driven sector.

Deteriorating security and resources shortages are national challenges that demand national responses. As an aviation nation, Australia needs its air power to be more than the sum of its parts. The first step in doing so is to think of it as a national air power enterprise.

Let me now unpack some of the ideas in his presentation and augment some of them.

I would start by turning to one of the RAAF's key focus, namely on the need to create a more agile force able to operate across bases in Australia, notably Northern Australia. Last year John Blackburn and I interviewed the then Air Commander Australia, Air Marshal Goldie about his thinking with regard to being able to do this.

Goldie commented: *We are developing concepts about how we will do command and control on a more geographic basis. This builds on our history with Darwin and Tindal to a certain extent, although technology has widened that scale to be a truly continental distributed control concept.*

We already a familiar with how an air asset like the Wedgetail can take over the C2 of an air battle when communications are cut to the CAOC, but we don't have a great understanding of how that works from a geographic basing perspective. What authorities to move aircraft, people and other assets are vested in local area Commanders that would be resilient to degradation in communications from the theatre commander – or JFACC?[7] *We need to focus on how we can design our force to manoeuvre effectively using our own territory as the chessboard.*

But to do this, the RAAF needs a civilian structure that can allow for this to happen. Where is the fuel? Where are the means to move the fuel on a distributed chessboard? Where is the personnel to support distributed logistical support for a distributed RAAF?

Without a well thought out civilian support structure, force distribution is challenging and maybe not doable. Normally when we talk about force projection, we are talking about going somewhere else to confront an adversary. For the RAAF, force projection is what they do to operate domestically.

The aviation enterprise which McInnes is calling for is part of the kind of mobilization which is crucial if defence is credible not just for Australia but more generally for liberal democracies. But shaping a feasible mobilization effort needs to start with the kind of enterprise thinking which McInnes provides, namely, to find ways to leverage domestic capabilities that are NOT part of the ADF but could be mobilized in case of crisis.

And in his presentation, he makes an important point when referring to the Chinese Communist Party and not to China. The CCP running China is the threat; it is not some benign nation state called China.

But I want to close by focusing on one very key point which McInnes underscored, which is the following:

In a speech in May, the Chief of Air Force said the Air Force was exploring autonomous air systems to build a "national ecosystem that can rapidly scale production of uncrewed systems." That is promising but surely we can be doing more than exploring. Australian industry has a demonstrated competence and comparative advantage in a field that is advancing rapidly around the globe.

There clearly needs to be a wider view of how to include autonomous systems in the force, and not simply thinking in terms of Air Force using autonomous air systems and the Navy using autonomous maritime systems as additive plus ups defined in terms of what manned systems currently do. It is about joint and combined uncrewed and crewed operating together across the battlespace.

It really is the broader notion of a combined arms paradigm or ecosystem if you wish whereby there is a collaborative relationship between autonomous, unmanned and crewed systems. If we conceptualize the new systems as simply fitting into what crewed systems do we will wait too long to use them and not understand what they can and can not do.

Above all, we need to move from experimentation to putting then in the hands of the warfighters as they augment the force in being and allow it to become the transformative force without waiting for the future force.

11
THE SPACE DIMENSION

Some time ago – more than a decade – I worked with Alan Dupas, the noted French space expert, on a project for a European space company on the future of space in 2020.[1] We focused on the key point that although a space company was most closely identified with launchers and satellites, the future was its engagement in the global information society.

Let me say that we were not greeted with cheers and love. Rather the major company we were dealing with shuddered at the thought that its "things" might be overshadowed by a product – data, communications and information. This of course puts the space company into competition with a range of providers of data, communications and information.

Space is an enabler of much which goes on in earth providing the nodes and networks of an information society. But space is costly, complex and governments are loath to invest more than they have to in such "esoteric" technology whose investments might cut into social spending or green energy or whatever the priority is for a sitting government.

This is certainly the case for Australia. Dr. Malcolm Davis at the seminar provided a compelling case for Australian space investments and acceleration of engagement in the space sector.

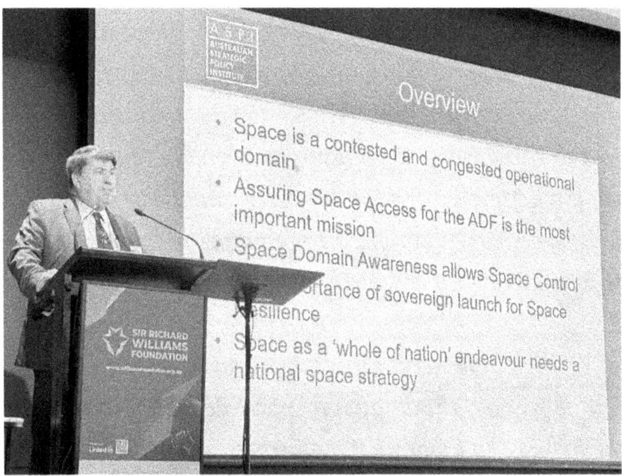

Dr. Davis speaking to the September 26, 2024 Sir Richard Williams Foundation seminar.

This is how presented and discussed this important subject:

Space is contested and congested. When we're talking about resilience, both being contested and congested are really becoming much more acute as a challenge.

Assuring space access for the ADF can be defined in different ways, but I would argue that it's not just about being able to use a foreign provider. It is also about sovereign space capabilities.

Space domain awareness allows space control. If you look at the national defence strategy and integrated investment program, it highlighted space domain awareness. They importantly made the point that space control is an important task for the ADF..

We can't have assured access to space if we rely only on foreign launch providers to give us that capability. We need to prioritize our national space capabilities, including sovereign launch. We need to pursue space policy as a whole of nation endeavor.

We don't currently have that.

It was started by the previous government. Those efforts were canceled by the current government. I would argue that we need to restore a whole of nation space strategy.

Space is an operational domain in its own right... We're seeing in the arsenals of our adversaries counter space capabilities. And these capabilities do not apply only in hot war scenarios. They could also be used in terms of gray zone operations as well...

We need to think in terms of how we defend against what the Chinese call system destruction warfare or how they can utilize counter space capabilities along with cyber attack, electromagnetic operations, and kinetic operations to take down critical Information Infrastructure as quickly as possible...

Part of resilience is managing space traffic and that requires a new approach to how we think about space domain awareness, how we manage the increasing amount of material that's in orbit.

Space is increasingly competitive in the sense that it's no longer just the sole domain of the major powers. It is also about the activity of small to medium powers, including Australia, as well as commercial actors.

And space has become democratized through a combination of falling costs that are driven by new technologies which allows more states to do things in space than previously was considered possible or financially viable.

That means there is a greater possibility that you could get either non state actors, commercial actors or hostile state actors essentially using space in a way that's inimical to our interests.

But it also brings opportunities in the sense that more states like Australia can actually do things in space that previously were beyond our capabilities...

We're starting to think about space 3.0. Space 1.0 was the Apollo era of big space agencies and the activities were the taxpayer funded and government led.

Space 2.0 was the establishment and the emergence of commercial space activities which really transformed the space environment and global space activity,

Space 3.0 is that next step that beckons in the future. It's that opportu-

nity to do space-based industry and a manufacturing capability, a space based economy that exploits space resources and new environments such as lunar space.

We have to challenge the orthodox mindsets that I think currently exist within government which primarily thinks about space in terms of satellites and rockets and start thinking about how we can utilize space in radically new and different ways that generate prosperity and growth.

He then went on to discuss how adversarial actions in space (war in space) can bring down or dismantle space infrastructure and that this infrastructure is a key part of a functioning information system for Australia.

This meant that the Australian government needed to get out of any stoved-piped look at space and take a broader view which would include space policy in the whole of nation concept of defence.

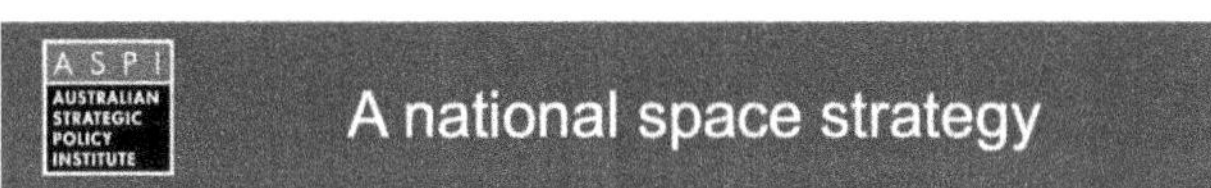

- If we are to be serious about assured access to space, and building space resilience, then this will demand government support the preparation of a **National Space Strategy** that integrates Defence's requirements with Commercial space and Civil space needs.
- A national space strategy – updated biennially to align with the NDS and IIP – should see Australia's future in space as **an enduring 'whole of nation' enterprise** that sits above party politics, to generate national prosperity and technological excellence.
- There needs to be sustained bipartisan support for growing the Space Sector across all dimensions – ground and space segment – space launch, satellite manufacture and space defence requirements.
- The preparation of a National Space Strategy that elevates the importance of space – and makes a clear case for space to the Australian people – **must be a priority for the next government after the election in 2025.**

A slide from Dr. Davis's presentation at the September 26, 2024 Sir Richard Williams Foundation seminar.

He then added: *The democratization of space technology means that space is no longer dominated purely by the major actors, so it's far more unpredictable as an operating environment. Increasingly, counter space technologies are moving in radically different ways and posing direct threats to space assets.*

For example, if you think back to the Cold War, there was no such thing as cyber warfare. Now we have the potential opportunity for cyber-attacks on satellites that can create scalable or reversible effects to disable or deny. And so suddenly, space weapons or space warfare or counter space capabilities become far more usable because it's in the interests of our adversaries to use them.

And I think that our adversaries recognize that space warfare and counter space capabilities can generate decisive strategic effect.

Space is critical for maintaining how we fight wars and how we undertake joint and integrated operations across multiple domains, but it's also vital for sustaining our information-based economies and societies...

Modern information-based societies depend on space capabilities to function, in particular through satellite communications, but also positioning, navigation and timing services. Everything that we do in a modern society from using information on our mobile phones, to our computers, to stock markets, logistics systems, all of that depends on the space capabilities.

That dependency will grow in the future, particularly as we get more and more reliant on processes of change associated with the Internet of Things and pursue the fourth industrial revolution. Such transitions demand that we have continued access to space.

Dr. Davis then went on the identify the various means of space attack and degradation which adversaries have already demonstrated.

And his point was clear – If Australia wants to protect its free and open society, if it wants to support a "rules-based" order which in my view is shrinking globally, how can you do so without an effective space engagement policy?

12

THE KEY ROLE OF DEFENCE INDUSTRY IN AUSTRALIA

Defence industry in Australia is obviously a key player in the ability for the government to find ways to enhance the ready force.

A panel of six industrial representatives discussed this challenge lead by Katherine Ziesing of the Sir Richard Williams Foundation as the facilitator.

The six panelists in order of providing presentations were as follows: Andrew Doyle, Director, Business Area Lead, Aeronautics from Lockheed Martin Aeronautics; Nick Leake, Head of Satellite and Space Systems, Optus; Derek Reinhardt, Director of Engineering and Operational Excellence, Northrop Grumman Australia; Dr. Brad Ferguson, Joint Battlespace Systems Technical Director, Raytheon Australia; Daniel Reinger, Engineering Manager MQ-28A Ghost Bat Combat Collaborative Aircraft (CCA), The Boeing Company; and Dr. Gary Eves, Principal Technology Officer of CAE Defence and Security.

The first speaker, Andrew Doyle, underscored how he viewed industry and its role in Australian defence resilience.

The critical enablers to growth in industry capacity--your experienced workforce, your facilities, and your capital equipment-- have lead times of years to establish the scale that Australia will potentially need.

We're already operating in a contested environment in terms of competing with other national priorities. With defence, where government is typically the owner, the operator and the regulator of defence systems, there's definitely a role for government to play in fostering that ecosystem for defence industry to be able to grow the scale and depth that Australia needs to be calling upon in the future.

To do this requires a well-considered investment strategy and in my view, a partnership with industry whereby industry can make investments that lead to capacity for them and capability for the ADF. Any disruption in investment ensures that capability will not be there for the ADF.

Andrew Doyle, Director, Business Area Lead, Aeronautics from Lockheed Martin Aeronautics.

Doyle put his assessment this way: *I will now talk about industry's role in building a resilient and scalable national defence ecosystem which starts with the basics of depth of industry presence and a close degree of integration between industry and defence. Industry needs to have the appropriate degree of insight into defence plans and capability and preparedness requirements for defence to leverage the additional mass and scale that industry can bring.*

The key to success is communication and close partnerships to ensure that we're getting alignment of resources and actions to best affect where industry can contribute to the operational viability of the integrated force...

The second panelist was Nick Leake. He underscored that the ADF cannot operate effectively without secure C2 and ISR, and that in today's world this means secure access to space and to satellites. He noted that Optus currently operates three geo satellites, one of which carries defence payloads.

Leake then highlighted the coming of the Optus 11 satellite which he indicated would be the first software-defined satellite in the region. He underscored: *With these new spacecrafts you will have fundamental intelligence on board, and you will be able to configure that satellite in terms of its capacity and where you're actually pointing that capacity.*

Putting a chip on board the spacecraft obviously then opens up further issues with cyber security, because you're putting intelligence in space, and you have to protect that asset the best you can.

He then highlighted Optus working on in-orbit servicing which will allow the satellite service life to be extended as fuel tanks are replaced on the satellite in orbit.

He next discussed the LEO revolution which is obviously a significant transformer of the space satellite eco system but indicated that Optus worked with partners to leverage LEO constellations and to shape an adaptive network.

Leake highlighted the importance of Dr. Davis's presentation on the need for enhanced sovereign space capability and indicated that Optus was a key part of any such effort.

The third presenter was Derek Reinhardt from Northrop Grumman Australia. His focus was on their work in enhancing the efficacy of sustainment in support of the ready force. NG is involved in the sustainment of the KC-30, the C-27J, the VIP fleet and the Triton.[1]

About two years ago, we were trying to bring all of our programs together and have them work in a more consistent way. To do so, we set about building a sustainment delivery model which was really intended initially to align our programs, but it's providing us interesting insights into the information that moves within a sustainment environment: the speed that that information needs to move, the decisions that hang off that information and how the enterprise combines to be able to do so.

Nick Leake, Head of Satellite and Space Systems, Optus.

He then when on to describe their creation in effect of a digital twin of the sustainment system. He went to argue that through this effort they have enhanced their ability to focus on the critical enablers for managing the information flows for decision making in the sustainment enterprise.

He noted: *What we've really learned to be successful, the architecture, the cyber-worthiness, and the whole concept of the data fabric is absolutely vital.*

When applied to the challenge of sustainment in a contested environment or contested logistics, this led him to the conclusion: *How the data fabric is architected is crucial for it to continue to operate.*

The fourth speaker was Dr. Brad Ferguson from Raytheon Australia. He certainly underscored the importance of enhancing the sense of urgency and speed to turn innovations into combat capabilities.

This is how he put it: *We need to adapt our architectures to support plugins for new capabilities, new technologies to support the rapid growth and leverage those technologies, everything from AI to quantum to hyper-*

sonics to directed energy to autonomy, these things will shape the future battlefield.

Derek Reinhardt, Director of Engineering and Operational Excellence, Northrop Grumman Australia.

He argued that the challenge and opportunity is to combine international cooperation with Australian delivery of capability. He then provided an example of this approach.

We started with the NASAMS system fielded in nine other countries, and then we built it from the ground up, integrating it with Australian innovation.

Working with over 30 companies across Australia, we integrated CEA radars, novel electrooptic infrared systems, new tactical data links to integrate with the Australian internet and military teams, integrating new missiles to leverage in service munitions, and ended up with the most capable short range ground-based air defence system in the world.

Some of those Australian innovations are now making their way back into the global community, supporting our allies and allowing for export opportunities.[2]

The fifth speaker was Daniel Reinger of The Boeing Company. His presentation was short and succinct and focused on a key area of developing and incorporating autonomous systems into the ADF.

This is what he highlighted: *What we need to focus on is building something that's built to adapt. That's in the wheelhouse of the collaborative combat aircraft, because if we don't build something that's adaptable, it will be obsolete before we even get it fielded.*

How do we evolve our thinking, so we actually build something that's adaptable?

Dr. Brad Ferguson, Joint Battlespace Systems Technical Director, Raytheon Australia.

The answer that we're coming to is embracing open mission system standards and embracing not just open architectures, but government defined open architectures.

What does that do?

It opens up a best of industry ecosystem where everyone can come to the party. It lowers the barrier of entry. When we talk about CCAs, we talk about machine autonomy, we talk about flight autonomy, we talk about crewed and uncrewed teaming.

It's simply too much for any one company to build the platform and then pull all of that together in a coherent manner.

By expanding the ecosystem and lowering the barrier to entry, you can get smaller and more companies that have niche skills into the effort.

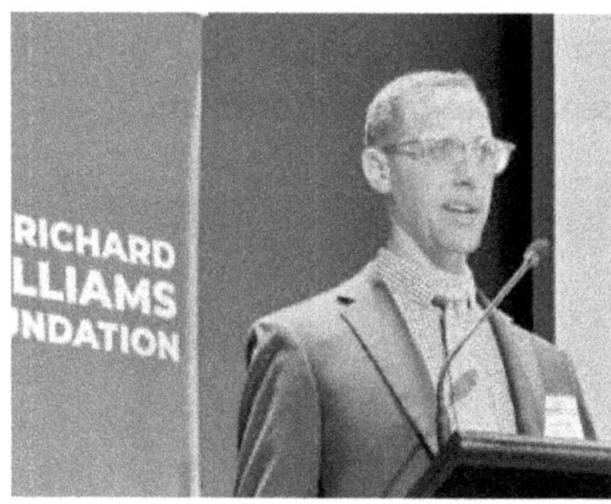

Daniel Reinger, Engineering Manager MQ-28A Ghost Bat Combat Collaborative Aircraft (CCA), The Boeing Company.

The final presentation was by Dr. Gary Eves from CAE Defence and Security who highlighted the growing importance in training and innovations in training to enhance the operational capability of the ready force.

He started by talking about the challenge for today's force in terms of training. *One of the things we need to understand is that the one size fits all approach just does not work for training. That requires a fundamental change in how we do things. What we are trying to do is evolve the capability of our young people to work with incredibly complex systems.*

Now it's not just a question of pure technical proficiency. They are decision makers. They're operating in a highly complex environment that requires dynamic decision making en masse, in real time, maybe without support.

He highlighted the importance of training for effective operation by teams in performing key tasks and missions which not only lead to mission success but to more rewarding experiences which are important in being to retain the personnel which you want and need for the organization.

He also underscored the importance of shaping effective ways for training in a coalition environment. This especially challenging because of different historical, linguistic and cultural experiences.

CAE has worked and is working on a variety of approaches to

succeed in the demanding training environments for the ready force operating in the new strategic environment.

Dr. Gary Eves, Principal Technology Officer of CAE Defence and Security.

There was a Q and A session after the panel presentations but the key focus was on the crucial need to reshape the partnership with government both for defence and commercial firms.

As one participant put it: *Don't try to design something in five years, because by the time you get it in five years, it's obsolete. We need an agile approach whereby we can build a capability, and then, over the years, we can add to that capability.*

Governments have to take some of the risk, and stop chucking the risk onto industry, because a lot of industries, particularly small business, will just walk away. If we have a shared risk and investment approach, we can have an agile model of delivering capability.

And the concept of "relational contracting" was introduced in a discussion of sustainment and support for the ready force, but perhaps has a wider application.

As one participant put it: *Relational contracting is an environment of defining how we work together, rather than defining specific technical requirements.*

Our best performing sustainment programs are those where you create the right relationship, you create the right dialog.

A shared situational awareness is created and shared understanding of who makes the right decision at the right time and with the right information which builds the trust that's needed for the desired outcome.

13
CONCLUSION TO THE SEMINAR

WGCDR Sally Knox functions as the moderator for the Sir Richard Williams Foundation seminars.

She provided a short but crisp conclusion to the seminar.

She underscored the following:

Several critical points emerged during the seminar highlighting the importance of balancing lethality with survivability and ensuring that our force is both operationally relevant and resilient in the face of accelerating threats.

Achieving combat mass across domains requires not only cutting edge technology, but also the ability to sustain and replace that mass over multiple rotations in high threat environments and responding swiftly and at scale across vast geographical areas.

This is key to ensuring our survivability and effectiveness in strategic relevant, strategically relevant timeframes.

The role of industry as well in mitigating sustainment and preparedness risk is pivotal. We must deepen industry collaboration to enhance our support base, workforce integration and capacity for innovation and building redundancy across critical nodes and domains.

Such collaboration is essential for surviving and operating in an increasingly complex threat environment.

14
INTERVIEWS SEPTEMBER-OCTOBER 2024

In this chapter, I have included a number of interviews which I conducted in Canberra in September and October 2024. Together these interviews provide a more detailed look at issues associated with ADF modernization by leading Australian strategists, industrialists and ADF officials.

The interviews addressed two major issues. The first is the overall strategic direction of Australian defence policy and the second is the question of how the ready force can be transformed rather than waiting for the new platforms to be delivered by the envisaged future force.

Andrew Carr highlighted that the emphasis on future platforms as the generator of future capabilities does not solve the problem of enhanced deterrence for Australia in and of themselves.

Dr. Stephan Frühling addressed the question of Australia and its working relationship with allies in terms of delivering credible deterrent capabilities in its region. Frühling's approach highlights to importance of linking any enhanced capabilities added to the ADF to a much broader defence and deterrence equation: what impact does it have in an alliance and partnership context?

How does it close gaps and seams in Pacific defence?

How does it enhance the credibility of the broader self-defence activities of Australia's partners in the region?

How does it mesh with and influence U.S. actions and policies in the region?

The remaining interviews looked at key aspects of being able to field a more effective ready force. A strategic shift which would do so is to emphasize a collaborative alliance effort in building up Australia as an anchor for munitions and missile production alliance-wide.

The inclusion of autonomous systems called for throughout the seminar presentations follows a different trajectory than does the acquisition of future platforms. Doing so effectively puts the ADF on a different course than one built around a future force dominated by manned platforms.

What does changing the ready force by leveraging an autonomous system revolution look like?

CAPABILITIES, COMMUNICATION AND CREDIBILITY: THE DETERRENCE EQUATION APPLIED TO AUSTRALIA

October 5, 2024

Recently, I met with Dr. Andrew Carr of Australian National University to discuss the way ahead for Australian defence policy.

There is much change underway for the ADF, but the focus has been largely upon capabilities, either adding them or dropping them.

For example, the government is committed to a new submarine for Australia, namely an SSN from the United States and then an AUKUS designed and built next generation submarine, and at the same time the government has cancelled the fourth squadron of the F-35 along with other programs to find ways to pay for this new capability.

But as Carr underscored: *Capability is not the same as deterrence. Deterrence is really a function of capability plus credible policy making and effective communication with adversaries and competitors before, during and after a crisis.*

What the 2024 National Defence Strategy has provided is largely a capability discussion but, in that review and the conversation ever since, the focus is upon deterrence, but in a very vague manner.

We need to unpack how the Australian government intends to deter adversarial actions against Australia, and simply adding a new capability, in this case SSN's, really does not begin to answer the question of what Australia is deterring.

This is what the 2024 NDS says about Australian deterrence policy: *Deterrence, is now Australia's primary strategic defence objective, and that our defence strategy seeks to deter any actions that could lead to conflict or coercion or direct military action against Australia.*

Carr then underscored that *this is new ground for Australia. We didn't do deterrence in the Cold War. It just wasn't part of how our defence strategy was organized.*

And we never have really had to do it on behalf of someone else.

And that means that our language is not focused on clarity with regard to threats or how we would deter or deflect them with the three C's: Capability, communication and credibility.

Carr argued that a new SSN capability for Australia clearly adds capability to the ADF. But its contribution to deterrence is not as clear.

He noted that much of the effort to redo the ADF and its infrastructure is in the North of Australia, but the new SSNs coming a decade down the road don't add to the deterrent value of changes in the North.

The partnership language in the region by the ADF and the government and working with the archipelagic nations like the Philippines and Indonesia could be part of a deterrence stated posture, but to date it has not been stated in such terms.

But this is also gets to the credibility question: what can the ADF and Australia as a nation credibly do to defend themselves against gray zone and other efforts of China in the region?

Where is Australia hoping to project credible power across Australia and into the region?

The government has clearly wanted to own all discourse on defence and has put a lid on the military but at a cost whereby there is little clarity about the focus of the deterrence effort.

The question then from Carr's analysis can be put bluntly: *What is Australia really concerned about from a deterrent perspective?*

The question clearly remains to clarify how the SSN capability relates to an answer to that question.

He mentioned the case of the Christmas and Cocos Islands. He argued: *There's been a long running public discussion about the fact that these islands have always been vulnerable to an enemy attack.*

The Australian Defence Force's capacity to retake them, if need be, would be incredibly difficult.

We're investing $100 million into the Cocos Islands for development, but the money alone is not enough.

The capability alone is not enough.

There needs credible communication that this is Australian sovereign territory, and we would pay the costs necessary to reclaim them.

Another example he provided was of the RAAF base Butterworth located in Malaysia. This is a legacy of the UK's days of empire, but now the context is very different. But how does this base fit into Australia's defence strategy and how is it incorporated into a deterrence strategy?

In short, although vagueness in articulating a deterrence strategy is important, unhinging it from capability enhancements means that a state cannot communicate an effective strategy. For Dr. Carr this a key challenge which the Australian government must meet head on and is not something which enhancing the capabilities of the ADF alone will solve.

AUSTRALIAN DEFENCE IN A BROADER ALLIANCE CONTEXT

October 31, 2024

When considering how the ADF is being reworked through the Defence Strategic Review for the long-term and the ADF's current force is is transforming, what is its broader impact on the region?

In my recent discussion with Professor Stephan Frühling, he focused on putting ADF and Australian actions in the larger regional context. Or in other words, what is the impact of Australian actions on broader alliance and partnership relationships in the region and beyond on deterrence of the authoritarian powers, notably China?

He started with the judgement that *the kinds of forces that we're planning for only makes sense as part of a wider multilateral deterrence and defence construct.*

The ADF that will result from the Defence Strategic Review will be focused on denial in Australia's approaches, but it cannot do so on its own against China.

But what is that broader construct?

For Frühling, the contribution of the ADF and Australia more generally is best understood in terms of what their contribution can be to "horizontal escalation" within a broader alliance and partner deterrence construct.

By which he means Australia's ability to work with allies and partners across the region to bolster their national defences and ability to resist Chinese pressure on the region and to work with partners beyond the region—especially in Europe—to encourage and facilitate their regular presence and signalling of support for a free and open Indi Pacific.

In particular when considering the ADF and Australian defence and security focus, one should consider how to close "gaps" in broader Indo-Pacific defence.

He underscored: *We need to start looking at what the seams are between what we're doing and what other countries are doing in our region. And I think a near term focus needs to be upon identifying those seams or gaps in capability and how we collectively can address them. If we don't address them, significant vulnerabilities are created which our adversaries will exploit.*

In a recent paper which he published with the Carnegie Endowment, he argued that the way ahead with regard to the core Australian alliance with the United States should follow such a path.[1]

This is what he wrote: *Both allies should consider practical cooperation in areas that reflect Australia's preparations for major war in its immediate neighborhood; that support multilateral deterrence by facilitating politically palatable horizontal, rather than vertical, escalation; and that help move force posture cooperation from enabling U.S. activities on Australian territory toward greater overall alignment of both countries' defence preparations.*

He then went on to identify five foci that both allies should consider to provide greater focus, purpose, and direction to force posture and structure cooperation.

He started by considering how the United States and Australia might broaden their consideration of what cooperation is in both their interests going forward. As he put it:

The United States and its Indo-Pacific allies have been grappling with how to best balance a rising China for years. Recently, they have often embraced the fuzzy concept of integrated deterrence. Yet they have not coalesced on a shared concept of deterrence and escalation that would direct how they think about the coherence and complementarity of their respective national force structure and posture developments.

But to do so it is better in his view to shape concrete steps ahead rather than vague policy declarations.

He underscored: *There is little appetite in Canberra to embark on the development of a document akin to NATO's Strategic Concept or the U.S.-Japan defence guidelines, which would only shove to foreground funda-*

mental disagreements, both between the allies and within Canberra itself, on deterrence and alliance strategy. But one way for both allies to work toward narrowing differences and identifying shared tenets is by aiming to say a little more each time their ministers meet at their regular summits.

Second, he highlighted focusing on mutual work on overlapping operational needs and challenges.

He provided several examples such as this one: *The defence of the U.S. West Coast and Australia's east and southeast present very similar challenges in terms of their geographic distance from adversary bases but increasing vulnerability to cruise missiles launched from SSNs or long-range bombers.*

While Australia is a similar size to the continental United States, the RAAF's roughly one hundred fast jets lacks the home-defence squadrons that the United States maintains through its National Guard. Australia's planned six NASAMS fire units will likely not just be adequate for the number of facilities that need protection, but also are ostensibly being acquired to defend forward-based land forces rather than, for example, irreplaceable submarine and naval bases in Sydney and Perth.

Joint examination of these issues, including drawing on the analytical work that underpins the U.S. homeland cruise missile defence program, could identify additional specific investments or preparations that would benefit both sides' wider operational objectives.

Third, the United States and Australia should also examine the overlap of their respective strategic and operational objectives in the South Pacific.

He gave this example: *In a major war, the southwest Pacific is significant for its geographic position along key sea lines of communication that would support the U.S. war effort, notably the lines between Hawaii and Townsville (where the great circle route passes through the Solomon Islands) and Townsville to Manus (and on to Guam or Palau), which passes east of the Papua New Guinea mainland.*

As in World War II, North Queensland would likely become the key staging area for U.S. operations from Australia, which is reflected in plans to move the combined logistics, sustainment, and maintenance enterprise from its initial location in Victoria to a future "Logistics Support Area" in Queensland.

U.S. convoys passing through the southwest Pacific would need protection against overt and covert Chinese lodgments in the islands. Australia would have an interest in playing a major role in this, not least because U.S. rules of engagement may well be more tolerant of collateral damage to South Pacific nations and their local shipping than Australia would be comfortable with.

Hence, examining the relationship between U.S. plans and concepts for strategic sea transport and force protection and Australia's increased focus on littoral operations in the same region would be a worthwhile area for joint planning and force posture cooperation. As convoys would also require protection against Chinese SSNs further into the central Pacific, there is scope for including New Zealand and France in broader discussions as well.

Fourth, he argued for encouraging European participation in the broader deterrence equation in the Pacific. Such participation has notably been increasing in recent years, and he believes such participation is important in policy signalling with regard to China in the region.

One could underscore the importance of Europe following similar policies as Australia with regard to China rather than providing Chinese policy end arounds. And the kind of policy signalling with European engagement can be important in supporting a broader Western deterrence policy towards China.

And in our discussion, he indicated a couple of examples of what could be of significant value in terms of European support for Pacific operations in case of conflict in the Pacific region. One would be setting up a tanker air bridge to support Australia and any forces deploying from Australia.

The second would be part of the broader arsenal of democracy. Here he mentioned what to my mind is an obvious way ahead, namely Australia buying into European developed and made missiles, which clearly has begun with Konigsberg. It seems clear that in a major war, Australia will be scrambling for any modern missiles it can get its hands on.

Fifth, he argued that UK involvement in Australian defence has a number of side benefits not often considered when talking in AUKUS terms.

This is how he put it: *One, often under appreciated, benefit of including the UK in AUKUS is that it has significant experience with the creation of multinational integrated military capability, including the kind of mixed crewing envisaged as part of the AUKUS optimal pathway.*

Such experience does not always transfer easily within the U.S. military and policy system and its relatively separate Euro-Atlantic and Indo-Pacific policy communities....

While British deployments to Australia in a crisis would be much smaller than those from the United States, involving the UK in, for example, discussions about the combined logistics, sustainment, and maintenance enterprise would enable deeper consideration of how to support and facilitate the deployment of other countries' forces.

From a purely Australian perspective, a greater understanding of the UK's decades-long and seemingly quite complex experience with U.S. nuclear and conventional bombers operating from its territory would also be useful in developing political and policy mechanisms to facilitate such deployments in Australia.

In short, Frühling's approach highlights the importance of linking any enhanced capabilities added to the ADF to a much broader defence and deterrence equation:

- What impact does it have in an alliance and partnership context?
- How does it close gaps and seams in Pacific defence?
- How does it enhance the credibility of the broader self-defence activities of Australia's partners in the region?

- How does it mesh with and influence U.S. actions and policies in the region?

RESHAPING AN ARSENAL OF DEMOCRACY: AUSTRALIA AS AN ANCHOR FOR ALLIED DEFENCE PRODUCITON

October 7, 2024

When one considers ways ahead in terms of allied Pacific defence, the question of supplies, magazine depth and logistics is a crucial and enduring challenge. Not only do Pacific forces need magazine depth with regard to munitions, but security of supply and capacity to move supplies in the region at the point of need is crucial.

The new Kongsberg plant in Australia is the glimmer of a new way to think about the way ahead for allied defence production for Pacific military operations.

A 5 September 2024 press release from the Australian government announced this development as follows:

The Albanese Government is further accelerating Australia's long-range strike capability through the acquisition of the Joint Strike Missile (JSM) from 2025. The Government and Norwegian company Kongsberg Defence & Aerospace have signed a $142 million contract to deliver the JSM for use by the Australian Defence Force.

The JSM is designed to allow the missile to fit into the F-35A Lightning II internal weapon bay, preserving the aircraft's stealth characteristics. With a range of more than 275 kilometers, the JSM's low-altitude sea-skimming flight profile helps it delay detection and engagement by a target's defence system.

The accelerated acquisition was announced today alongside Norway's Minister of Defence, Bjørn Arild Gram, at the opening of Kongsberg Defence Australia's new $25 million manufacturing facility in Mawson Lakes, South Australia. Kongsberg's new South Australian facility will have the capacity to employ up to 150 workers and will assemble launchers *for the Naval Strike Missile (NSM) using mostly Australian-manufactured components, creating 20 new local jobs.*

This announcement follows the Government's recent commitment to partner with Kongsberg Defence Australia for the construction of a new facility at the Newcastle Airport precinct capable of manufacturing and maintaining the NSM and JSM – creating more than 500 jobs in the construction phase and delivering almost $100 million in economic benefits to the local area. [2]

One should note that the JSM is enabled in significant part by the building an alliance-wide combat aircraft, the F-35. When working for Secretary Wynne from 2004-2008, the possibility of allied production of a common strike platforms enabled by a common missile was a key point we often underscored.[3]

As important as this development is, we need to think and invest strategically. The United States is no longer the arsenal of democracy as we saw in the Second World War or even the Cold War. If Australia is expecting the U.S. to produce weapons in depth for Australia, Australians are kidding themselves.

And the American forces simply do not have any magazine depth. And it is even worse that that. How does the U.S. transport weapons to the operational force with a limited logistical force? Not enough tankers, not enough airlifters and not enough ships for the Military Sealift Command are available for rapid and timely support and resupply.

And how do allies such as South Korea and Japan who use common missiles with Australia and the United States shape magazine depth in the region and enhance the security of supply?

I had a chance recently to discuss this challenge with my colleague Stephen Kuper of *Defence Connect*.[4] What we focused on was the opportunity for Australia to become an anchor for defence production in the munitions area including guided munitions but much expanded over the current concept.

Rather seeing Australia doing limited production for itself, the United States and Europe would expand their notion of security of supply beyond what they produce in their home markets for export. Rather, American and European investments would be made to jointly procure the magazine depth which allied forces need. This also would allow for a more secure location from which to move necessary support to deployed forces.

This is how he put it: *The U.S. and Australia joined by perhaps Japan and South Korea would shape a joint order book to shape a predictable and steady stream of orders which would allow plants in Australia to attract the investment they need to build out capacity.*

This might well attract investment from the larger investment funds because of predictable and steady production at scale from Australian plants for the ADF and allied forces. This would also lead to sorting through the logistical means necessary to move munitions to the partners and to their operational forces.

This is what I call "embedded logistics" whereby one designs the support center with modes of delivery built into the production construct.

The point can be put bluntly: our capabilities to support our forces with the munitions they need are simply not there. And the logistical capabilities to move the munitions we do build are inadequate.

Simply stockpiling weapons in Australia for U.S. forces will be too little and too late. We need a reconceptualization of the production-support-logistical system.

This is one way to do so.

MEETING THE CHALLENGES OF INNOVATIONS FOR THE READY FORCE

October 22, 2024

How can organisational design drive innovation within a military?

I spoke with John Blackburn during my September-October 2024 visit to Canberra about this challenge. He has dedicated much of his professional life to addressing such issues.

In our conversation, Blackburn argued that successful innovation within the operational force requires a combination of bottom-up initiatives, top-down design, and a tolerance for failure while attempting to innovate. He recalled the work of Admiral Cebrowski of the U.S. Navy on network-centric warfare, where he emphasised the importance of experimentation to drive innovation, while accepting failures as part of the learning process.

Blackburn stated that a well-defined CONOPS is the crucial link between bottom-up innovation and top-down design. He recalled his time in the Royal Australian Air Force during which he worked on such an approach for the RAAF's Plan Jericho.

He described the approach this way: *The theme of Plan Jericho's fifth-generation Air Force was "top-down design meets bottom-up innovation." The idea was to establish a concept of operations as a test idea, then trial it with bottom-up innovation without stifling creativity.*

The concept would be adapted based on these experiments because it remains largely theoretical until tested to see what works. Plan Jericho was founded on the principle of tolerance for failure, learning from experiments, and continuously improving the operational force.

He continued: *We were talking about a feedback loop. You start with a design, and bottom-up innovation helps refine it. You iterate the design, recognising what works and what doesn't. Another challenge is dealing with bureaucratic control. I was the mentor to the team and assisted with the design of the plan and its execution. We funded it initially at $15 million a year and started exploring innovative ideas in partnership with industry.*

After about a year, we identified several promising bottom-up innovations. My advice to the team was to hand off these plans to the running RAAF system, which excels at carrying out well-defined tasks. Once handed off, the Jericho team could then refocus on evaluating the CONOPS, identifying what went wrong, and considering alternative approaches.

Unfortunately, the Australian Department of Defence subsequently moved towards a centralised innovation model focused on long-term platform acquisition. Centralising innovation is a surefire way to kill innovation as it becomes bureaucratised and limits the ability to rapidly redesign and innovate within the operational force. The issue is one of risk tolerance and focusing on platforms rather than the operational effects they create.

Blackburn expressed concern that the current approach of the Australian Department of Defence appears to emphasise platform acquisition without a coherent top-down design. There is a lack of what a 2015 Defence review called "program-level" design to integrate the myriad of projects in the capability investment plan. This is a major problem, for example, in acquiring an Integrated Air and Missile Defence (IAMD) system.

Blackburn argued: *We are buying billions of dollars' worth of equipment for an IAMD system without a coherent system-level design.*

Blackburn further argued that CONOPS-driven design aligns with another key requirement: the need for a debate on a National Security Strategy to guide defence strategy. Australia does not have a National Security Strategy; it has a defence strategy without broader guidance that integrates all elements of national power.

In summary, Blackburn emphasised that innovation in defence requires a clear concept of operations that connects top-down design with bottom-up experimentation. This approach helps create an adaptive, ready force rather than a static, platform-centric force that struggles to keep pace with evolving threats and requirements.

AUTONOMOUS SYSTEMS FOR THE READY FORCE: IT IS ABOUT THE CON-OPS

October 8, 2024

Throughout the day of the Sir Richard Williams seminar on 26 September 2024, speakers highlighted the importance of unmanned and autonomous systems enhancing the capabilities of the ready force, rather than seeing them as part of some future force.

But such systems are best understood as providing payloads for the kill web rather than as platforms which are unmanned or uncrewed. By characterizing them this way, one can envisage how the ready force will get their hands on these systems and use them.

And for this to happen, the new systems need to shift from being platforms as science projects to payloads which the warriors can integrate into their concepts of operations. That means as well, the enthusiastic science fiction writers about autonomous systems will have to temper their expectations.

I had a chance to talk about this challenge with Vice Admiral (Retired) Tim Barrett after the seminar. As former Chief of Navy he emphasized that when he was in office he argued that these systems needed to be put into the hands of warfighters rather than isolating them as science projects.

We talked about one such system, namely the Ghost Shark, which is built by Anduril and is being tested and evaluated for use in the near to mid-term.

But as we discussed, for such an ISR capability – for the payloads most useful in the near term by UUVs will clearly be part of the SA network for the Royal Australian Navy and its partners – to be effective it will have to be used by those who manage underwater SA and operations in the RAN.

This is how Barrett put it: *The reality is that for autonomous systems to come into the current force, they need to be well practiced at the operational end – promotion of their adoption is a behavioral piece. New systems need to be in the hands of warfighters to ensure that these systems make the current force more agile and take actions that are effective in their application.*

Operational success is still about the application of force in the mind of the person responsible for delivering it. It's about forcing them to think of operational success by whatever means they have available to them, and then having the courage to take those actions.

You are not necessarily after disruptive change in process, but disruption in the effect. In some cases you don't want disruption in the efficiency of the process of operations. But, you want to be able to cause a disruption that has an effect on your adversary.

With regard to the Ghost Shark, to fully achieve its potential, it has to quickly enter the operational world of those who are managing the underwater warfare space throughout the regions of our interests.

To be effective as a disruptive technology, it will need to contribute to the operational effects being sought by those managing the undersea domain; in tactical terms this means it has to be of benefit to those managing the water column. It could generate strategic consequences but not simply because of its technology but in the way this it is used to produce disruptive operational effects.

A successful water space management process is key to being able to determine where your adversary is, or, more importantly, where it isn't, so that you can put the right forces in the right place.

Bureaucracies don't necessarily think like that. Operators absolutely do so, because it's their day-to-day business, and they're in the practice of only putting in harm's way those things that need to be there to affect a disruption to the enemy's operations.

The disruptive effects that a Ghost Shark can produce should be determined by those who actively manage the battle space, the undersea battle space, rather than someone who's programming from afar and doing so in complete isolation from the rest of the water space management concepts of operations.

THE CHANGING LITTORAL OPERATIONAL CONTEXT: WHAT ROLE FOR MARITIME UNMANNED AND REMOTE SYSTEMS?

October 11, 2024

I had a chance to follow up with Jennifer Parker on her excellent presentation at the 26 September 2024 Sir Richard Williams seminar which focused on the evolving threat environments in the littorals and insights to be gained from operations in the Black Sea, Red Sea and the Philippine's Sea.

She argued in that presentation that new capabilities, notably USVs and UAVs used by the Ukrainians and the Houthis posed new challenges to capital ships in the littorals. And that capital ships clearly can still be effective but ongoing modernization of their defensive systems in the new context on an ongoing basis was critical.

In our discussion, she underscored that the threat from land systems was of enhanced range, and new threats posed by unmanned maritime systems introduced additional threats to capital ships as well.

What this meant for her was the absolute importance of ongoing modernization of the combat systems aboard surface ships. Rather than viewing updates as occurring in long periods of block upgrades, there needs to be an ability to weave in upgrades based on the rapid evolution of offensive threat systems from various operational theaters.

It is crucial to have very credible threat information from a diversity of deployments by the Royal Australian Navy and its allies, and to be able to weave that information into ongoing upgrade efforts of combat systems.

She referred in the seminar presentation to her recent paper she published in the *Australian Naval Review* concerning the USVs used in Ukrainian operations against the Russian Black Sea Fleet. In that article, she made the important point that these were NOT autonomous systems but remotely piloted ones.

She called for precision in analysis of any lessons to be learned from Black Sea operations. This is how she put it in her article:

The USVs employed by Ukraine are what this article would consider light (less than 5 tonnes), remotely controlled attack USVs. This is important, as the lessons from the Black Sea are not necessarily scalable to large USVs (greater than 1,600 tonnes) or even small USVs such as the United States's Sea Hunter (approximately 500 tonnes).

The USVs employed by Ukraine are not autonomous, which is often misunderstood. The ease at which Ukrainian USVs have targeted some surface ships through remote control using electro-optical and infrared sensors cannot be scaled to supposed developments in autonomy.

The lessons from Ukraine's employment of light attack USVs are different from the lessons you would learn regarding employment of light USVs such as a sail drone, Bluebottle or similar that are being trialed for intelligence and surveillance capabilities.[5]

The significant effects which the Ukrainians have had using their USVs in the Black Sea clearly showed that countries without much of a capital ship force could still pack a punch in the littorals.

But actions in the Black Sea also demonstrated that the Russian fleet was not prepared with proper defence systems and training against such a threat. And this means that viability for capital ships operating in such waters clearly need to have such systems and such training.

But then we focused on the significant question of how capital ships can be combined with autonomous ones in shaping new combined arms capabilities at sea for a maritime force. It simply makes little sense to send our most advanced destroyers into harm's way into the littorals to fight the Houthis.

She expressed her concern that both the U.S. Navy's LUSV program and its RAN counterpart were not perhaps the most credible addition. It is in effect an optionally manned arsenal ship with launch tubes to add to the strike capability of the fleet.

Frankly, I am very skeptical that such a program is going to change the nature of the fleet anytime soon given the concern one has over firing authorities and defensive systems. Precisely the concern which Parker raised regarding the need to upgrade defensive systems on capital ships would apply to an LUSV as well.

We then focused on how UUVs and USVs as autonomous systems could enhance the manned maritime fleet.

In my view, the way ahead is shaping a maritime kill web force. And UUVs and USVs are not platforms but really payloads encased in a carrying system. Or put in other terms, what specific mission capabilities do they add to the manned force? Or how can they work as part of a combined arms maritime operation?

Parker added that it has been overstated what small numbers of these systems can bring to the fight. And clearly, from an operational perspective these systems need to be put in the hands of operators to determine how to use what additive or replacement capabilities these systems can provide.

The software nature of AI maritime systems requires operators and the company/government team writing the code to work closely together in the evolution of desired and effective capability. By taking a combined arms perspective rather than a manned-unmanned teaming perspective, the focus is specifically on what a specific payload carried by an autonomous carrier contributes to a specific operation and operational capability.

But then we closed by focusing on a key organizational barrier to this route. The RAN and the U.S. Navy are focused on distributed maritime operations, but hierarchies in the defence bureaucracy have made it increasingly difficult for local decisions to be made by the operational force.

In an interview I did with a senior U.S. Admiral last year, he put this challenge bluntly: "When I do mission rehearsals, I find gaps that need to be filled. We can identify gap fillers we should be able to buy to make the distributed fleet more lethal and survivable."

But of course. he cannot do so given the policy and acquisition hierarchy blocking the innovation which the ready force can discover and implement readily.

If this is not remedied, the promise of autonomous systems for rapidly improving the lethality and survivability of the ready force will not be fulfilled.

THE PATH TO LEVERAGING AUTONOMOUS SYSTEMS

October 25, 2024

Recently, I met with Marcus Hellyer, the well-known Australian defence strategist, to talk about the challenge of getting autonomous systems into the operational force. He has been working with UUVs and I have been working with USVs, and both systems face challenges of getting them into the hands of the operators and thereby entering the operational force.

We discussed why this has been so difficult.

Hellyer started with the straightforward point that autonomous systems or uncrewed platforms are not simply a new variant of manned platforms and should not be understood in these terms.

You can't look at autonomous systems as simply an unmanned version of a traditional platform. Everyone says that but I don't think they really think through what that means.

And what it does mean is you don't want it to do everything that a traditional platform does with the autonomous system, because if you try and design it to do so, it's going to be just as complex as a manned system.

This in turn means that is going to take just as long to design and it's going to cost just as much as a manned platform. There won't be savings in terms of time, money and people.

In other words, the key point to underscore is this: Start simple, design autonomous systems to do one thing, and once they can do one thing effectively, and you work from there as the operators use them and input their demands into this process.

The point bluntly put is that autonomous systems and their development must be operator driven development, not the least of which because they are payloads managed by a software governed platform.

And we turn to a major theme highlighted in the 26 September 2024 Sir Richard Williams Foundation which was how to add combat mass to the force. And as Hellyer underscored: *There is no way to achieve this goal except by incorporating autonomous systems into the combat force.*

He then focused on the term often used in describing working with autonomous systems, namely manned-unmanned teaming. He warned that this effort can lead to the focus on how legacy or new capital ships can onboard such systems and be part of that ship's combat systems. He argued that such an understanding would undercut the use of the new systems, because they have the potential to operate independently of capital ships as part of a kill web rather than needing a mother ship to carry and launch them.

Certainly, capital ships can and will be redesigned to incorporate an ability to launch and recover autonomous air and maritime systems, but they are payloads which can operate separately from the capital ship fleet or in conjunction.

What is entailed is a naval version of the fifth-generation air revolution. Fifth generation air platforms essentially operate in a kill web of shooters and sensors and are not designed as the primary weapons carrier in the fleet.

This means that as autonomous systems deliver payloads to the ready force as part of the operational kill web, naval warfare will change. But if a government already has formed a legacy PLUS approach to its capital shipbuilding shipbuilding program, how would it adapt to the significant change which autonomous systems – air and maritime – will deliver maritime operations?

I have talked with the Danes who are designing their new generation frigates along these lines and who are working with Singapore in designing such a new generation support ship as well.

Will Australia re-calibrate its ship building approach as autonomous systems enter the fleet and alter maritime operations?

In the seminar, the Commander of the Royal Australian Navy, RADM Christopher Smith, provided insights which could suggest the possibility. For example, he said:

Future multistage systems will incorporate adaptability and interchangeability that goes beyond the power of carrier centric forces, as the diminished size and the crewing requirements make this capability available and affordable additions to even frigate and destroyer sized platforms.

But more than just uncrewed systems, future multistage forces will leverage the joint capabilities of distributed land and air assets throughout the maritime environment. The ability of future ground and air forces to contribute essential kinetic, non-kinetic effects to multi domain strike missions in the maritime environment will be essential to improve the economy of effort and generate sufficient precision and firepower at the decisive points.

A clearer statement of a focus on the maritime kill web could not be made, and in such a force the shift is to the payloads and combined effects one can create not simply on the platforms organized in legacy task forces.

APPENDIX 1: THE SEPTEMBER 26, 2024 PROGRAM

Enhancing and Accelerating the Integrated Force: An Operational Perspective

AIM

The aim of the September 2024 seminar is to examine the enhancement and acceleration of the Integrated Force from an operational perspective.

The seminar seeks to identify the most significant factors which impact the Whole of Australian Government, Defence and industry, as well as international partners in a multi-domain context with an increasingly complex set of threats and operational risks.

BACKGROUND

In a transition from an agenda focused on the 5th Generation Australian Defence Force, the 2024 seminars examine contemporary Defence challenges, gaps, and opportunities in the context of the current round of Defence and defence industry reviews across the time periods highlighted in the 2024 National Defence strategy:

Now until 2025 – the Enhanced Force-in-Being will focus on immediate enhancements that can be made to the current force.

2026 to 2030 – the Objective Integrated Force will see the accelerated acquisition of critical capabilities.

2031 and beyond – the Future Integrated Force will see the delivery of an ADF that is fit for purpose across all domains and enablers.

Four strategic themes provide a focus for the Sir Richard Williams Foundation; they cover mass and depth, agility, industry involvement, and redundancy.

The intent of the themes is to provide an objective consideration of those measures which balance combat effectiveness with an increasing demand for efficiency in a fiscally constrained national security environment.

BUILDING COMBAT MASS AND DEPTH ACROSS DOMAINS:

How can new technology be applied to the extant force structure to enhance a focussed force and achieve critical mass and depth?

The ability to mass power, plus sustain and replace it includes consideration of force protection and generation requirements for sustained operations across multiple rotations in a high threat environment where success is by no means assured.

GENERATING TEMPO ACROSS DOMAINS:

How can the ADF respond at speed and scale across a vast geographical area in strategically relevant timeframes and remain survivable?

ENHANCING INDUSTRY CAPABILITY AND THE NATIONAL SUPPORT BASE:

How can industry be more involved in planning to mitigate key sustainment and preparedness risks not least in relation to cost, and the broader demand for increased capacity and workforce integration?

SURVIVE TO OPERATE:

How does the ADF build redundancy to sustain integrated operations across multiple domains and critical nodes in the face of an increasingly complex and lethal threat?

Beyond the ADF there is the broader need for an objective assessment of industry competencies, workforce preparedness, and national will. Moreover, a central question emerges around whether the nation is ready for the realities of a future war?

SEMINAR AGENDA

The September seminar will focus on enhancements to the focussed force, and the gaps and risks emerging from the NDS and Integrated Investment Program.

The seminar will seek to support the transition to a Future Integrated Force that is fit for purpose across all domains and enablers that considers the rapidly accelerating threat and the need to prioritise investment in capability which is operationally relevant, survivable, lethal, sustainable, and delivered with the lowest political risk.

The proposed industry focus is: How can industry be more involved in planning to mitigate key sustainment and preparedness risks in the integrated force, and introduce new technology to add mass and depth plus the ability to sustain and replace it?

Cutting across each of the strategic themes are policy, process, technology, infrastructure, and workforce considerations including, but not limited to:

- Operational Risk Management: Balancing Lethality and Survivability
- Workforce: Management, Automation, and Development
- Training System Effectiveness & Mission Rehearsal
- Future Technologies and Collaborative Combat Systems
- Accelerating Capability: Minimum Viable Capability and Minimal Viable Product
- Preparedness, Basing, Logistics & Supply Chains

- Assuring Value for Money in Force Development

Enhancing and Accelerating the Integrated Force: An Operational Perspective

26 September 2024, National Gallery of Australia

Program

Time	Topic	Speakers
0800-0830	Registration and breakfast	
0830-0835	Welcoming Remarks	VADM (Rtd) Tim Barrett AO, CSC Sir Richard Williams Foundation
0835-0840	Introduction and MC	WGCDR Sally Knox Sir Richard Williams Foundation
0840-0900	Politics and Defence in the AUKUS countries - the agenda in the 2020s	Peter Jennings AO, PSM Director Strategic Analysis Australia
0900-0920	Building Resilience in the Space Domain	Dr Malcolm Davis Australian Strategic Policy Institute
0920-0940	The contested maritime domain: Challenges for an integrated force?	Jen Parker, Expert Associate, National Security College Australian National University
0940-1000	National Air Power Enterprise	Chris McInnes Executive Director, Air Power Institute

1000-1030	Break – Morning Tea	
1030-1130	Panel Discussion: Building Combat Mass	AIRMSHL (Retd) Darren Goldie AM, CSC– Facilitator RADM Christopher Smith AM, CSM, RAN Commander Australian Fleet AVM Glen Braz AM, CSC, DSM Air Commander Australia
1130-1230	Industry: Enhancing Technology - panel	Katherine Ziesing - Facilitator Andrew Doyle AM, CSC, Director, Business Area Lead, Aeronautics, Lockheed Martin Aeronautics Nick Leake, Head of Satellite and Space Systems, Optus Derek Reinhardt, Director Engineering and Operational Excellence, Northrop Grumman Australia Dr Brad Ferguson, Joint Battlespace Systems Technical Director, Raytheon Australia Daniel Reiniger, Engineering Manager MQ-28A Ghost Bat Combat Collaborative Aircraft, The Boeing Company Dr Gary Eves, Principal Technology Officer, CAE Defense and Security
1230-1330	Lunch	
1330-1350	Video - PACAF Perspective	Gen. Kevin B. Schneider Commander, Pacific Air Forces
1350-1410	Joint Capabilities Perspective	LTGEN Susan Coyle, AM, CSC, DSM Chief of Joint Capabilities
1410-1430	Enhancing and Accelerating the Integrated Force – an RAF Perspective	AVM Mark Flewin CBE MA BEng RAF AOC 1 Group, Royal Air Force
1430-1450	The Integrated Force – an ASD Perspective	Phil Winzenberg Deputy Director General – Signals Intelligence and Effects, Australian Signals Directorate
1450-1510	Chief of Air Force Perspective	AIRMSHL Stephen Chappell DSC, CSC, OAM Chief of Air Force
1510-1530	Formal Close	Sir Richard Williams Foundation

APPENDIX 2: THE PERSPECTIVE OF LT GENERAL STUART SIMON: PRESENTATION TO LAND FORCES 24

Recently, the Chief of the Australian Army provided his perspective on the way ahead for the Australian Army.

He did this appropriately at the recently held Land Forces conference held at Melbourne, Australia.

The conference was challenged by protestors whose performance and presence reminded one of how internal conflict within the liberal democracies is dovetailing into the broader challenge to "the rules based order" being posed by the authoritarian powers externally.

Lt General Simon Stuart is a thoughtful military leader in these very challenging times and his speech provided insights into how he conceptualizes the context and the approach to shaping a way ahead for the Army within the context of the ADF as a joint force.

He entitled his speech: **'The Human Face of Battle and the State of the Army Profession'**

That speech follows:

We gather here today in what are widely agreed to be the most uncertain times in several generations.

The 'end of history' has proven to be little more than a holiday, as once again we find ourselves in a world defined by great power competition.

Large-scale inter-state war again blights the world stage. For some, aggression clearly remains a viable tool of statecraft. The current bloody conflicts in Ukraine and the Middle East demonstrate the grim verity of Clausewitz's insight into war's enduring nature.

As a professional soldier and a student of history, I spend much of my time focused on the tension between war's enduring nature and it's ever changing character.

It is plainly evident that the acceleration in technology is changing how warfare is prosecuted, just as it is in every other facet of life. Those with access to advanced technology can now see more, sense more, and strike faster ... with more accuracy and greater lethality.

Some argue that these changes are so profound as to be revolutionary.

But amid the seductive tones and tints of the true chameleon we recognise the timeless challenges of land warfare, challenges that have not fundamentally changed in more than two millennia.

Thucydides himself would understand the nature of the fighting in Eastern Ukraine, despite its prosecution by novel weaponry.

Foch, Monash or Von Paulus might nod knowingly at the calculus that inspired the recent Ukrainian thrust into Kursk.

The armies fighting in Ukraine and the Middle East are finding that technology is anything but revolutionary, when set against a highly-motivated enemy fighting in complex terrain.

As the American scholar Stephen Biddle pithily observed, current warfare looks more like the trench impasse of the Great War than any fictional Star Wars.

For this reason, and so many others, I am especially delighted to welcome Doctor Jack Watling and General David Berger as our keynote speakers.

Doctor Watling's work in providing scholarly, yet accessible, analysis of the war in Ukraine has only enhanced his global reputation as one of the foremost contemporary thinkers and writers about land warfare.

General Berger brings a lifetime of dedicated service that encapsulates warfare and military adaptation at the very highest levels. His personal experience and applied scholarship are second to none.

Theirs are the voices we would do well to heed if we are to understand the nature and character of war in our time.

Our theme today is "The Human Face of Battle".

I chose this difficult topic for the Symposium very deliberately.

It is easy in this modern age to become fascinated by the allure of technology. Allure, however, becomes a siren's song when these technologies offer to sanitise war: to make political violence remote, risk-free, quick and clean.

Of course such images are appealing. Who wouldn't seek a future where we are untouched by the brutality of war? But the evidence to support such a proposition is absent.

One need only glance at Aleppo, Mariupol, Mosul or Marawi to conclude that Thomas Hobbes's state of nature ... a life that is 'solitary, poor, nasty, brutish and short' ... remains the reality for many and is, uncomfortably, far closer than we might wish.

In the Guns of August, her seminal study of the start of the First World War, Barbara Tuchman reminds us that 'the human heart is the starting point of all matters pertaining to war'.

So, our focus today is about whether warfare remains an intrinsically human activity, and what that means for those of us who are accountable to build and steward armies.

How do we best weave together the human elements of fighting power – the intellectual and the moral – with the physical elements, to become more than the sum of their parts?

My opening view is that we should approach this topic with a great deal of humility, and respect for war's unpredictable nature.

In Homer's Odyssey, Odysseus tied himself to the mast of his ship before he allowed himself to listen to the siren's song, lest he lead his ship off course. We ought to take similar measures.

It is a genuine privilege to engage with this audience, and one should always seek to use such an opportunity with clear purpose and intention.

I therefore intend to focus my remarks on one aspect of the 'human face of battle' that I believe to be of fundamental importance if we are to successfully meet the challenges of our time.

And that is the state of the army profession.

Less than 18 months ago, the Australian Government received the Defence Strategic Review: an articulation of policy that will be as generationally impactful for the ADF as the 1987 Defence of Australia white paper that has shaped the last four decades ... and the service of all of us in this room today.

The DSR – and the National Defence Strategy that emanated from it – rightly concluded that Australia's current strategic circumstances are significantly changed, that great power competition increases the risk of conflict, and that the ADF is not fully fit to meet these challenges.

The DSR concluded that, for the first time in nearly 80 years, we must go back to fundamentals ... to take a 'first principles' approach. Our Army has fully embraced and applied this approach – setting us on a transformational path.

There is not a soldier, a team or a formation in the Australian Army that is untouched as we optimise for littoral manoeuver and deliver a long-range fires capability.

In less than a year and a half we have achieved much. We have re-written our capstone concepts and TTPs to reflect our littoral future, and have re-orientated our command and control to reflect how we would fight.

We have transformed the Army Training Enterprise to ensure we can generate resilient, well-trained soldiers and cohesive teams.

We have re-focused the roles of our combat brigades, and began consolidating them in northern Australia.

We have re-defined the Division as our unit of action, and have reorganised the Army accordingly.

And perhaps most notably for many in this audience, the Government has moved out to lay the capital foundations for our transformation.Recent decisions on long-range fires, watercraft, the combined arms fighting system, land C4, and battlefield aviation mean that the physical component of our fighting power is increasingly assured.

I am immensely proud of what our soldiers have achieved, and my confidence is buoyed daily by their energy, spirit of innovation, and sheer determination.

As we swiftly traverse the foothills and early summits it is clear that genuine transformation implies, indeed demands, that we consider all aspects and dimensions of our Army: to go beyond the physical component of fighting power, and to address the fundamentals that the DSR challenges us to consider.

We must examine in breadth, width and depth our intellectual and moral foundations. For history tells us it is here that the margin between victory and defeat will be found.

So, I intend to lay the foundations today for an assessment of the state of the Army profession.

But where to start? As ever, we must understand where we have come from in order to imagine our future.

The Australian Army is a comparatively young army, and the strength of our professional foundations have waxed and waned throughout the course of our history.

The Great War broke out only 13 years after Federation, and represented a rite of passage for the newly-formed Commonwealth of Australia into the global community of nations.

The Australian Army was at the centre of this baptism of fire. It was due in no small part to Australia's military commitment that the Commonwealth was party to the Versailles Peace Talks, and was established as a founding member of the League of Nations.

Our Army's status as a national institution was secured by the soldiers of the First Australian Imperial Force, and 'Anzac' became part of the foundation story of modern Australia.

The cost was very high. Over 60 thousand Australians were killed in that war from a population of less than 5 million.

Rapid mobilisation was followed by rapid demobilisation, and exacerbated by the financial challenges of the Great Depression.

The period between the two World Wars was a difficult time for those drawn to soldiering.

Opportunities for service were few, so much so that many of Australia's best and brightest Duntroon graduates pursued the call of their profession elsewhere in the Commonwealth.

Money was tight, our geography seemed impregnable, and the Royal Navy was supreme on the global commons. Australia's apparent need for a professional army declined accordingly.

But the war to end all wars, did not. Just twenty years later a second world war darkened the horizon. Initially the fight was, once again, distant, but by 1942 it reached our shores.

For the first and only time in our history, so far, our homeland was directly attacked by the conventional forces of a hostile nation state.

The need for an Army became urgent, even existential.

History tells us that we were ill-prepared for the brutal three-year fight that followed in the Pacific. It is a testament to the small cadre of professional Australian soldiers, the flexibility of the Citizen Militia Forces, and sheer national will that we were able to build the amphibious and littoral Army needed to fight and win on a battlefront that at one stage stretched from Bougainville to Borneo.

The Army not only survived in the Pacific, but was victorious, and its standing as a national institution and a fighting force were enhanced.

Bonds were formed with our American ally, indelibly sealed in battle and blood, that have stood the test for over eighty years.

It was only in the aftermath of the trauma of the Second World War that the need for a standing, professional Army became self-evident to our Nation.

And so in 1947 the Australian Regular Army was formed. This is where service in our Army achieved the status of the profession that we recognise today. A profession provides a service to society that society cannot provide for itself; maintains and advances a distinctive body of knowledge; and is expected and trusted by the society it serves, to self-regulate.

It is this moment that defines the Army that I have the privilege to lead today, an Army built not just on the Dardanelles Campaign, but also on the foundations of victory in the Pacific.

Throughout our history, our Nation's demand signal for a professional

Army has ebbed and peaked. A simple pattern is discernible. In times of regional or global conflict, the demand has risen, be it in the World Wars, in Korea, or in Vietnam.

But in times of peace ... in what we might call the 'inter-war' periods ... the need has ebbed.

The health of our profession has perhaps followed a similar pattern. Our professional foundations have been strongest in times of war, and especially when our Nation and its interests have been directly threatened.

This is reflected in the names of those great Australians who continue to personify our Army's profession, in names like Bridges, Blamey, Monash, Vivian Bullwinkel, Charlie Green, and Harry Smith.

But in the inter-war periods our Army has routinely contracted, reducing to a small professional cadre that has striven to maintain the foundations of soldiering.

This cadre have done well throughout our history but this contraction has invariably incurred a debt that had to be paid in the early stages of the next conflict. It has taken time to focus the profession, to adapt it to the character of warfare it was about to face. To steel it once more for the rigors of combat.

This debt was paid in raising and expanding the 2nd AIF, in the re-training of the 6th and 7th Divisions for the war in the Pacific, and in the adaptation of the profession for Vietnam, East Timor, Iraq and Afghanistan.

Today we do not enjoy the luxury of time. To quote the Defence Strategic Review, 'in the contemporary strategic era, we cannot rely on geography or warning time'. A conflict in our region would constitute an immediate threat to Australia and its interests.

So, we must ensure our profession is fit for today's purpose ... that we are ready to 'fight tonight'. This is our responsibility ... indeed our obligation.

As one of my predecessors, Sir Henry Wells, adroitly put it in 1957, we must 'avoid the situation where soldiers have to be killed to learn'.

But where do we begin in assessing the state of our profession? Fortunately, there is a solid foundation of theory upon which we may rely.

In 1957 the American theorist Samuel Huntington codified the foundations of the profession of arms in his timeless work The Soldier and the State.

He was joined in 1962 by the British soldier-scholar General Sir John Hackett, whose Trinity College lectures captured the soul and the art of the profession better than any practitioner since.

There are others, but Huntington and Hackett are foundational. It is from their work that I draw the three pillars of the profession that will form the basis of our review.

The first and most important pillar is that we provide a service to society that society cannot provide for itself. In particular, in Huntington's terms, the Army specialises in the mastery of violence on the land for socially determined ends.

This is a responsibility enshrined in the very foundations of our Nation, articulated in the text of our Constitution and instruments pursuant to it.

But it is an authority that we must never take for granted. Our permission to apply violence on behalf of society relies on trust as much as it relies on legislation, if not more.

This is one of many reasons why I have selected 'trust' as the central strategic priority for our Army today.

Trust and social license are explicitly linked: lose one, and we lose the other.

Hackett captures the essence of this first pillar in his artful articulation of the 'contract of unlimited liability'.The professional soldier accepts that they may be required to forfeit their life at the behest of the nation. Their liability is unlimited. This is a solemn commitment indeed.

But all contracts have reciprocal obligations. If soldiers are to accept a contract of unlimited liability, then it is both our Army and our Nation's obligation to honour that commitment. The final report of the Royal Commission into Defence and Veteran Suicide, published not three days ago, should give us pause to reflect on how well we are meeting that obligation for all whom serve.

It is clear to me that strengthening the Army profession must be at the centre of our response to the Royal Commission.

The second pillar is our professional body of knowledge, our ability to teach it, and the sufficiency of both to meet the demands of the future.

Here we are quite well postured. The Australian Army has developed over generations a genuinely world-class training and education system. We are one of the foremost leadership training institutions in the Nation, with over 25,000 leaders amongst our ranks, from Lance Corporal to Lieutenant General.

But even these foundations will be insufficient for the scale of the challenges we face. Professor Sir Michael Howard once wrote that 'no matter how clearly one thinks, it is impossible to anticipate precisely the character of future conflict. The key is to not be so far off the mark that it becomes impossible to adjust once that character is revealed'.

I am not assured that we yet have the doctrine and learning systems in place to hard-wire adaptation into our profession.

The war in Ukraine reminds us that most conflicts are in fact a battle to adapt ... an action, reaction, counter-action fight for a decisive edge. Speed is vital ... the side that can adapt fastest gains the advantage.

Would we win the battle for adaptation in our region? My view is that this is by no means certain, and that we must do more.

Experience tells me that the study of the liberal arts is vital to our profession. In his excellent book 'The Face of Battle', for which this Symposium is named, John Keegan explains (and I paraphrase) that 'what battles have in common is human: the behaviour of people struggling to reconcile their instinct for self-preservation, their sense of honour and the achievement of some aim over which others are ready to kill them.

The study of battle is therefore always a study of fear, of courage, of leadership and obedience.

Battle is an historical subject, whose nature and trend of development can only be understood down a long, historical perspective'.

Are we doing enough to teach our young professionals about this perspective; about both the art and the science of war? The Australian Defence Force Academy, for example, provides an excellent education to our young officers, but the study of the classical theories and practise of warfare remains an elective choice.

Not many take up this choice. Only around 50 of the 900 annual graduates of that excellent institution study Australian military history: well less than 10%. Most leave having never heard the names of Clausewitz or Jomini, having never studied in depth an operation or campaign.

We might ask ourselves if this is the best foundation for them to flourish in our profession.

The third and final pillar rests on our capacity for professional self-regulation.

Militaries are routinely given the authority to regulate good order and discipline through military codes of justice and discipline acts in national legislation.

But this isn't what I seek to examine. Rather I am focused on the intangible forces of self-regulation: the virtue-ethic of the Australian Army, our philosophy as a fighting force, and a culture that urges us to hold ourselves to the highest professional standards.

It should not surprise anyone in this audience when I observe that we must do more to reflect on the sufficiency of our professional standards in recent conflicts. What I refer to as the 'Long Shadow of Afghanistan' will continue to shape the context for our Army for years to come.

From mission command to command accountability, we need to understand what worked, what didn't, and how we can add steel to our professional foundations to prevent them fracturing in the crucible of combat.

So, to draw these ideas together, the theories of Huntington, Hackett and their peers will be our guide. We will draw much from their wisdom. But equally we will not be dogmatic. It is already evident that there are some lessons of sixty years ago that are likely best consigned to history.

Huntington, for example, considered professionalism to be synonymous with 'officership'. For the Australian Army in the 21st Century, such a *restriction is unacceptable. Every soldier must be considered, and consider themselves, to be part of our profession. To accept anything less will be to risk failure from the outset.*

In closing, I offer my sincere thanks for your patience. I do not underestimate the challenge I lay out for our Army today.

Throughout our history we have conducted many reviews of key elements of our profession: the Regular Officer Development Committee of 1978, the Project OPERA reviews of 1998, and the 2016 Ryan Review to name but a few.

But as far as I can ascertain, this will be the first time since 1947 that we have attempted a wholesale, holistic review of our profession. It will be a hard road, but it is a necessary one for us to traverse.

I contend that our profession must be fundamental to our Army: a 'first principal' that underpins and shapes all others. We must consider it, understand it, invest in it ... but above all we must believe in it.

It is my obligation as the accountable steward of the Australian Army to set us on this path. Today is the first of these steps. I will return in the coming months to speak again on the state of the Army profession, and in the next year I will clarify our priorities, and what it is we need to do to realise the potential of the Army profession.

Today, however, I leave you to reflect on the words of General Douglas MacArthur to the graduating class of the US Military Academy, West Point in 1962 – the very same year that Hackett was lecturing at Trinity College.

'Through all this welter of change and development your mission remains fixed, determined, inviolable.

It is to win our wars. Everything else in your professional career is but a corollary to this vital dedication. All other public purposes, all other public projects, all other public needs, great or small, will find others for their accomplishment.

But you are the ones trained to fight. Yours is the profession of arms, the will to win, the sure knowledge that in war there is no substitute for victory. That if you lose, the nation will be destroyed ... that the very obsession of your public service must be Duty, Honor, Country.' Thank you.[1]

APPENDIX 3: THE IMPACT OF THE AUSTRALIAN POLITICAL SYSTEM ON NATIONAL SECURITY

By John Blackburn and Anne Borzycki

Australia has a lengthy history of seeking protection from great and powerful western saviours due to a long-held fear of invasion from "the north."

Stunned by the withdrawal of British Forces following the surrender of Singapore in WWII, it turned its attention to another large and powerful ally, the United States.

This dependence grew stronger following the UK's withdrawal from 'East of Suez,' which was precipitated by the Suez Crisis in the late 1950s. The domino theory of the 1960s, which predicted the spread of communism in South-east Asia, fuelled Australian public concern and was used for political gain.

Australia relied on the United States as its security guarantor, initially through the 1951 ANZUS treaty. A treaty that, with the rupture of relations between the United States and New Zealand, no longer exists in full force but is still trumpeted by Australian politicians whenever it is judged beneficial.

The bilateral U.S.-Australia alliance is the most strategically significant relationship, despite the announcement of the AUKUS partnership in September 2023 which implied some degree of relevance for the UK in the region.

National security is based on capabilities, competence, and partnerships; complacency is a major hindrance to that security.

Is Australia, in fact, secure in 2024, or is Australia a complacent nation adrift in the South Pacific?

POLITICS, GEOPOLITICS AND THE DEFENCE OF AUSTRALIA – THE 21ST CENTURY CONTEXT

On 14 September 2001, at a press conference at the Parliament House of Australia, Prime Minister John Howard invoked the Australia, New Zealand, and United States Security Treaty (ANZUS) for the first and only time, following the terrorist attacks on 11 September 2001 in the United States.[1]

John Howard had always been a conventional, if somewhat conservative, supporter of the Australian-U.S. alliance. Neither a champion nor a critic. That would all change after 9/11.

In 2006 Professor Emeritus Robert Manne wrote that "from this moment Howard worked conscientiously to create a new vision of the future: of an Australia deeply integrated – strategically, economically, socially and culturally – into the most formidable empire the world has ever seen ... the U.S.."[2]

Possibly a slightly hyperbolic assessment in 2006, but one that ultimately became Australia's reality.

Prime Minister Kevin Rudd's *Plan for Defence,* released during the election campaign of 2007, noted that the Labor approach to Defence "... is built around three fundamental pillars – [first] Australia's alliance with the United States ... [which] is fundamental to our national security and our long-term strategic interests."[3]

At the launch of the Defence White Paper (DWP 2009) on 2 May 2009, Kevin Rudd stressed that Australia's alliance with the United States would remain the bedrock on which the country's national security is built: "The document judges that the United States will remain the most powerful and influential military actor out to 2030."[4]

The DWP 2009 announced the need to upgrade and enhance maritime capabilities as a consequence of an assessment of the changing strategic environment.

Specifically, "by the mid-2030s, we will have a heavier and more potent maritime force. In the case of the submarine force, the Government takes the view that our future strategic circumstances necessitate a substantially expanded submarine fleet of 12 boats in order to sustain a force at sea large enough in a crisis or conflict to be able to defend our approaches (including at considerable distance from Australia, if necessary), protect and support other ADF assets, and undertake certain strategic missions where the stealth and other operating characteristics of highly-capable advanced submarines would be crucial."[5]

The phrase "*to sustain a force at sea large enough in a crisis or conflict to be able to defend our approaches (including at great distance from Australia, if necessary*" outlines a picture for the future submarine operating far from home, protecting our sea lanes and contributing to Asian regional security, and an unidentified threat, In 2009 Australia's political leadership was hesitant and discreet in overtly naming 'China' as the target of defence planning and capability expansions.

Over the period in which the Australian Labor Party (ALP) was in government (2007 – 2013), despite the internal fractures of two leadership coups, the commitment of Australian Defence Force (ADF) personnel to both the Middle East and Afghanistan continued almost unquestioned. The relationship with the U.S. remained foundational, and indeed, deepened.

In 2011, then Prime Minister Julia Gillard and President Barack Obama, announced two new force posture initiatives that would significantly enhance defence cooperation between Australia and the United States.

The media release said that "Starting next year, Australia will welcome the deployment of U.S. Marines to Darwin and Northern Australia, for around six months at a time, where they will conduct exercises and training on a rotational basis with the Australian Defence Force ... The intent in the coming years is to establish a rotational presence of up to a 2,500-person Marine Air Ground Task Force."[6]

The leaders also agreed to closer cooperation between the Royal Australian Air Force and the U.S. Air Force that would result in increased rotations of U.S. aircraft through northern Australia.

The announcement prompted rapid, and at times, strident, criticism from the Chinese government. *The New York Times* noted that the deployment / basing of U.S. Marines in Australia "prompted a sharp response from Beijing, which accused Mr. Obama of escalating military tensions in the region ..."[7]

In May 2013, just months before losing the federal election in September, the ALP released the Defence White Paper 2013.

This White Paper saw a significant shift in the language used in reference to China. From being regarded as a growing economic power in the region with whom meaningful dialogue and engagement was important, in conjunction with allies and friends, we now had a comforting paragraph reassuring Australians that there was no need to choose between the U.S. and China.

Australia's national security and economic reality appeared to be well balanced between the strategic rivals with whom we were connected.[8]

When Labor lost power to the Liberal/National Party coalition in September 2013, Tony Abbott, the new Prime Minister, took over the submarine program with a few proposals of his own. Prime Minister Abbott became infamous for his 'captain's pick' moments in which he made decisions without consultation with colleagues, advisers, experts or indeed, anyone.

Abbott became convinced that the Japanese Soryu-class submarine was precisely what Australia needed. Media reporting during 2014 widely reflected this position.[9] Local shipbuilding businesses were angered at the probable loss of jobs and commercial opportunities.

In early 2014, the newly elected Abbot Liberal/National Party government actively sought to deepen economic ties with China through a free-trade agreement, at the same time as balancing strong criticism of Chinese actions in the East China Sea.[10]

Prime Minister Tony Abbott famously described the relationship with China at that time as one based on 'fear and greed'.[11] Fear of China's growth, wealth and ambitions, but also greedy for the trade this growth enabled and the economic benefits that flowed to Australia. A variation of the 'we don't need to choose sides' position of the previous ALP government.

By February 2015, following a series of internal party tensions and leadership threats and challenges, Prime Minister Abbott announced that a Competitive Evaluation Process would be undertaken as part of the Future Submarine acquisition strategy. This evaluation process involved France (DCNS), Germany (ThyssenKrupp Marine Systems—TKMS) and the Government of Japan as potential international design partners.[12]

In September 2015, Prime Minister Turnbull overthrew Tony Abbott, in yet another bitter internal party coup. The free-trade agreement, sought by Abbott, was finally signed in December 2015, and the praise was heaped upon President Xi by Australian politicians at a formal dinner to mark the occasion in the Great Hall at Parliament House Canberra.[13]

Not long after the signing of the Free Trade Agreement, the Government released Defence White Paper 2016 (DWP 2016) which acknowledged "The relationship between the United States and China is likely to be characterised by a mixture of cooperation and competition depending on where and how their interests intersect."[14]

The DWP 2016 verified that the Future Submarine Program would be a rolling acquisition program.

On 26 April 2016 Prime Minister Malcolm Turnbull declared

DCNS (now Naval Group) of France as the preferred international design partner for Australia's future submarine. The successful design was the Shortfin Barracuda Block 1A conventional submarine, which was based on the French Barracuda nuclear powered submarine.

Australia had evidently not learned the lessons from prior defence acquisitions which involved taking a proven design by a foreign manufacturer, and then extensively changing it to meet perceived special Australian requirements.

On 30 September 2016 a contract worth between $450 and $500 million was signed between the Australian Government and Naval Group for the 'design and mobilisation' of Australia's 12 Future Submarines. This included the development of Adelaide's Osborne North submarine facility for this purpose.[15]

Regionally, the honeymoon period that followed the free trade agreement with China did not last long. Prime Minister Turnbull ignited the ire of the Chinese government in December 2017 when the federal government passed the Foreign Influence Transparency Scheme against the backdrop of a series of high-profile scandals involving CCP influence in Australian politics.

Though the law itself was not targeted at any one country in particular, Prime Minister Turnbull's use of the politically-charged phrase "the Australian people stand up," and his open criticism of CCP influence in Australian politics generated diplomatic blowback in Beijing.[16]

Then, in August 2018, Australia prohibited Chinese giants Huawei and ZTE from participating in building Australia's 5G networks. A tit-for-tat escalation of trade actions/sanctions followed which only worsened after April 2020 when the new Prime Minister, Scott Morrison (who overthrew Malcolm Turnbull in yet another bitter internal party coup), endorsed an independent investigation into the origins of the COVID-19 pandemic, infuriating the Chinese officials.

The 2020 Defence Strategic Update (DSU) signed by both the

Prime Minister and Minister for Defence, gave no hint of what was to come for the submarine project, just one year later.

The problems with the submarine project were a poorly kept secret around Canberra – within Defence and amongst successive Prime Ministers, Ministers, and industry. Indeed, there was minimal reference to the project in the DSU; merely a brief mention that the "Government will continue to deliver this nationally significant program of investment in ships and submarines..."[17]

And what was said about China?

Yet another step-change in the language used, for example: "While still unlikely, the prospect of high-intensity military conflict in the Indo-Pacific is less remote than at the time of the 2016 Defence White Paper, including high-intensity military conflict between the United States and China ..."[18]

Then Prime Minister Morrison further worsened international relationships when, in September 2021, he announced the cancellation of the French submarine project after entering into a new, secretly negotiated, agreement with the United Kingdom and the United States.

With a single stroke Morrison aggrieved both China and France.

The French Foreign Minister Jean-Yves Le Drian called the trilateral submarine deal evidence of 'duplicity', 'treachery' and a 'stab in the back.' He also criticised Morrison's lack of candour. France recalled its ambassadors to the United States and Australia.[19] French President Emmanuel Macron said Prime Minister Morrison lied to him about his intentions with the deal, and that his trust in Australia had been deeply damaged.

This came a day after U.S. President Joe Biden expressed deep concerns about the handling of the secret plan to dump Naval Group from the future submarine project, labelling it as "clumsy" and "not done with a lot of grace".[20]

Many Australians would have a similar opinion of Morrison's honesty and clumsiness, especially after many of his activities as Prime Minister were made public following his defeat in the 2022 federal election. Of course, many Australians would also have vivid memories of France's less-than-trustworthy actions in the Pacific region over previous decades.[21]

THE EMERGENCE OF AUKUS

AUKUS, the trilateral security partnership for the Indo-Pacific region between Australia, the United Kingdom (UK), and the United States (U.S.) was announced on 15 September 2021.

The decision by then Prime Minister Morrison, was a complete surprise to the Australian people. In the March 2023 *Australia in the World* podcast, the late Alan Gyngell, former Director-General of the Office of National Assessments, a leading Australian foreign policy adviser and academic, was quoted as saying that "the most surprising thing about this announcement of the largest project ever undertaken by the Commonwealth of Australia remains the fact that there has been no formal articulation of the reasons for the decision. No report, no speech to parliament, no speech at all, other than the sales patter from successive governments: 'China is more assertive, the rules-¬based order is under threat, nuclear submarines are just what Australia needs."[22]

It's as if AUKUS, once proclaimed by the Liberal Prime Minister, became sacrosanct, with the then-opposition leader, now Labor Prime Minister, Albanese agreeing to the collaboration in just hours. There has been no political debate on this critical matter for Australia's security.

Now, it appears to be nearly treasonous to question the decision's merits; citizens debating the subject were effectively mocked by the Chief of Navy, Vice Admiral Hammond, who urged Australians to ignore 'hand-wringing' doubters of the AUKUS pact.[23]

Have political talking points spread to the highest levels of the Australian Defence Force (ADF)?

Whilst the 'U.S.' part of AUKUS could be seen as a logical fit, the value of the 'UK' part was more difficult to justify. Assuming that

AUKUS planning was already underway, and was a little over a year from being announced, what did the 2020 Strategic Update say about Australia's relationship with the United Kingdom?

In the 68-page document there was not one mention of the United Kingdom. Yet by September 2021, Prime Minister Morrison was quoted as describing AUKUS as a "forever partnership for a new time between the oldest and most trusted of friends."[24]

Recently, the Australian Defence Minister that the "UK has a much greater presence in the Indo-Pacific than we have seen in a very long time".[25]

But apparently this is news in Britain. The UK Parliament's *UK Defence and the Indo-Pacific – Report Summary,* dated 24 October 2023, stated: "Although we welcome the progress made in the region, we reject the notion that the 'tilt' has been 'achieved' from a Defence perspective.

"With only a modest presence compared to allies, little to no fighting force in the region, and little by way of regular activity, UK Defence's tilt to the Indo-Pacific is far from being achieved. If we aspire to play any significant role in the Indo-Pacific this would need a major commitment of cash, equipment and personnel, or potentially rebalancing existing resources. The UK Government's future strategy for the Indo-Pacific is still unclear."[26]

So much for a "forever partnership" with the UK in the region.

Political commentators, national security experts, academics and the media in general were still trying to come to terms with what AUKUS practically meant for Australia when, in April 2022, Prime Minister Morrison announced that a federal election would take place in May 2022.

The Liberal/National Party coalition had styled themselves as the bastions of security and the 'liberal' ALP were frequently derided as 'soft.' In the period leading up to the election, the ALP had little political wiggle-room to change stance on AUKUS or risk the usual 'soft on security' barbs being levelled at them.

At the end of the day, after nearly a decade of 'conservative' government, the Liberal/National Party was defeated, and the ALP was returned to government in May 2022 – with AUKUS steadfastly a part of their policy platform as well.

To emphasise this reluctance to 'rock-the-boat' so to say, the Australian public policy journal, *Pearls and Irritations*, observed: "It is understandable that the leadership team heading towards an election was determined not to be wedged on defence policy ... [but] why would an intelligent forward-looking government fail to fully assess the implications of secret international negotiations made by an unreliable former prime minister in the dying days of a discredited government?

"Why is our apparently very capable leadership team so timid about questioning the terms and range of preconditions for the AUKUS agreement?"[27]

AUKUS firmly intact, the new Labor government, under the leadership of Prime Minister Albanese, made immediate attempts to repair / renew the relationship with China – primarily as it related to trade.

There was an urgent political and economic need to get the export market back on track. And there were early breakthroughs in late 2022 as both the Foreign and Trade Ministers were able to meet with Chinese government officials to discuss avenues for lifting sanctions / tariffs and reopening channels of communication.[28]

Just one month after the election and during his visit to India, the Defence Minister/Deputy Prime Minister said in an interview with Australia's ABC: "... for India and Australia, China is our largest trading partner. For India and Australia, China is our biggest security anxiety. We're both trying to reconcile those things, which is not an easy problem to solve."[29]

Australia's perennial paradox, and a juggling act that is only getting more challenging. Australia's defence policy at a macro level rarely moves too dramatically from one government to the next, nonetheless, within the first 100 days of office, the Albanese government commissioned the independent Defence Strategic Review (DSR) 2023.

Within the body of the DSR, the language regarding China assumes a more ardent tone: "Intense China-United States competition is the defining feature of our region and our time. Major power competition in our region has the potential to threaten our interests, including the potential for conflict ... China's military build-up is now the largest and most ambitious of any country since the end of the Second World War ...

"This build-up is occurring without transparency or reassurance to the Indo-Pacific region of China's strategic intent ...China's assertion of sovereignty over the South China Sea threatens the global rules-based order in the Indo-Pacific in a way that adversely impacts Australia's national interests. China is also engaged in strategic competition in Australia's near neighbourhood."[30]

The relationship with China is a genuinely baffling example of cognitive dissonance. According to the Department of Foreign Affairs and Trade, China is Australia's largest two-way trading partner, accounting for 26 per cent of our goods and services trade with the world in Financial Year 2022-23.[31] Yet China remains the significant factor in Australia's defence planning, and posturing.

In the years since John Howard committed Australia to the United States, the relationship has been turbocharged to the point where the 'dependency' (rather than interdependence) could impact Australia's independence and sovereignty.

The DSR 2023 reflects this turbocharging: "Our Alliance with the United States will remain central to Australia's security and strategy. The United States will become even more important in the coming decades. Defence should pursue greater advanced scientific, technological, and industrial cooperation in the Alliance, as well as increased United States rotational force posture in Australia, including with submarines. Contrary to some public analysis, our Alliance with the United States is becoming even more important to Australia."[32]

There is an expression that was coined during the Cold War, by then-Prime Minister Menzies, to define what was virtually a canonical faith in Australia's international outlook – that Britain and the U.S. formed a protective umbrella for the country in a threatening Asia.[33]

The 'great and powerful friends' would be the protectors. From Federation until World War 2 it was the UK, and since then, the U.S.

- And while solid, reliable alliances are fundamental to global and regional stability, at what point does an 'alliance' become an 'addiction', an unhealthy dependency?
- Has this historic reliance weakened Australia?
- Has it given rise to complacency?
- Has it stifled creativity and innovation?
- Has it denied the nation a shared narrative?
- Has it created a gilded cage from which there is no escape?

With AUKUS, Australia has become a loyal partner to two empires – one in an apparent political civil war and the other of little relevance to the Asia Pacific region, according to its own Parliament, as discussed previously.

Indeed, the partnership may, in the end, constrain sovereign thinking, capability and sovereignty given the degree of compliance required with U.S. International Traffic of Arms Regulations and the Australian Governments inventive definition of 'sovereign' in the Defence Industry Policy.

The Policy defines foreign primes that have operations in Australia as 'Australian' and as 'sovereign' as actual Australian companies, which have majority Australian ownership and controls and headquarters in Australia.[34]

THE CURRENT APPROACH

On 24 April 2023 the Australian Labor Government released the Defence Strategic Review (DSR), which examined whether Australia possessed the required defence capabilities, posture, and preparation to best defend itself and its interests in the current strategic environment.

The DSR concluded that the ADF was "not fully fit for purpose" in the face of escalating security challenges. This occurred despite an investment of around A$695 billion in defence from 2007 to 2022 by both the Labor and Liberal Coalition governments.

How could this have occurred? What does the present Labor government plan to do about it?

The DSR stated that "Australia's strategic circumstances and the risks we face are now radically different - the U.S. is no longer the unipolar leader of the Indo-Pacific. Major power competition in the region has the potential to threaten our interests, including the potential for conflict. China's military build-up is now the largest and most ambitious of any country since the end of WW2 ... This necessitates a managed, but nevertheless focused, sense of urgency. It is clear that a business-as-usual approach is not appropriate."[35]

However, Australia's strategic situation is more complex than depicted in the DSR. Australians are currently dealing with concurrent, and in some cases existential, concerns.

These include climate change and the urgent need to reduce emissions, rising global and regional security risks, a global pandemic with long-term societal and economic consequences, a global energy transformation in which we lag behind the developed world, and a global market model that has resulted in reduced resilience due to a lowest price, just-in-time philosophy.

Increased levels of internal instability, societal unrest, and a lack of trust in government and institutions among our AUKUS partners raise concerns about the possibility of similar tendencies occurring in Australia.

In 2021, our Institute of Integrated Economic Research – Australia_produced a report on Australia's national resilience titled *Australia – A Complacent Nation ... Our reactions are too little, too late, and too short-sighted.*[36]

It highlighted Australia's lack of resilience in the face of current emerging threats and risks. Little progress has been made in the intervening three years.

Whilst the DSR did identify that a central component of deterrence for national defence is resilience, listing 12 examples of critical resilience requirements, there is no indication of how Government intends to address them; listing them is not acting.

In each case there needs to be a risk assessment before a strategy and plan can be developed. For example, given national concerns regarding the state of our energy systems and massive import dependencies, it is sobering to realise that the last time Australia had a National Energy Security Assessment, was in 2011. There appears to be no coherent plan.

The DSR was a largely bottom-up, platform-focussed review and plan, delivered incrementally over a period of 10 months, following its initial announcement, and devoid of any National Security Strategy or Defence Strategy to provide context and guidance. A Defence Strategy was delivered in April 2024, but there is still no elusive National Security Strategy.

It is also difficult to discern how the DSR has taken account of the numerous failures to implement previous reviews, as well as what lessons were learned from the actions that were only partially implemented.

The Government has stated that "Realising the ambition of the Review will require a whole-of-government effort, coupled with significant financial commitment and major reform."[37]

Despite the aspirational phrasing and free use of the terms "soon" and "sooner" in numerous political statements, there is no significant new financing for DSR implementation this decade.

The Australian Defence Department has often struggled to manage complicated change programmes, but now they are expected to succeed with this large and under-resourced reform programme. Reorganisations, reallocation of resources and cancellation / changes in projects are assumed to be all that is necessary in the near term.

The failure to resource the DSR adequately could mean that our deployable military operational capability will be less at the end of this decade than it is today and this right when the threat envisaged by the review could appear.

Since 2000 there have been six Chiefs of Defence who have been directed by 13 Defence Ministers (two of whom only served 271 and 272 days respectively), 13 Assistant Ministers, and 12 Ministers for Defence Industry, Defence Materiel, and Defence Personnel. A parade of 38 Defence related Ministers over 23-years, each with a party-political agenda, and often with little real understanding of National Security or Defence.

This august group tasked a total nine Defence related White Papers, and nine Defence reform, restructure, review, and transformation programs. Were any lessons learned during this 24-year journey? Apparently very little, given the findings of the 2023 DSR.

THE SUBMARINE SAGA

In 2009 the Australian Government announced that by the mid-2030s, we would have a heavier and more potent maritime force. Our future strategic circumstances necessitated a substantially expanded submarine fleet of 12 boats in order to sustain a force at sea large enough in a crisis or conflict.

Over the intervening 15 years, successive Governments have identified Japanese, then French and finally a mixed fleet of used U.S. submarines, and a yet to be designed UK submarine type with delivery of the latter to occur in the 2040s.

This, a decade after they are needed according to DWP 2009, and a decade after the heightened threat of conflict in the region could materialise, according to the DSR 2023.

Of particular interest, given growing public scepticism regarding the actual feasibility of the AUKUS submarine plans, is the view of the current opposition leader, Peter Dutton, who was the Defence Minister at the time of the AUKUS announcement in 2021.

In March 2023, Dutton warned against acquiring a future submarine fleet from the UK, stating that advice he received as defence minister before the last election was that the British companies involved in the UK's submarine production had no extra capacity to support an Australian program and insisting that American Virginia-class submarines remained the best and most practical option for Australia.[38]

A pity it took him 18 months to speak up but that's politics.

This incredible US/UK hybrid submarine fleet package is thoroughly examined in Professor Hugh White's superb essay "Dead in the Water," which was published in the *Australian Foreign Affairs Journal* in February 2024.[39]

He explains the decision's assumptions, the significant likelihood that the plan will be disrupted, and how deploying the proposed nuclear submarines in the future will have little or no impact the military balance in Asia or on Australia's security.

He discussed the capacity differences between conventional and nuclear-powered submarines, stating that while a nuclear-powered submarine may be equivalent to two conventional boats, it costs five times as much.

Professor White summed up the AUKUS deal as follows: "AUKUS also gave Morrison a high-profile defence initiative which provided a national security focus for the 2022 election campaign, and presented a chance to wedge Labor if they opposed it. Labor dodged that bullet by fully supporting AUKUS, partly from a pathological fear of a political brawl with the Coalition on any national security issue, but also because many of the party's leaders believe as fervently in supporting America against China as the Coalition does.

"It is a bipartisanship built upon political opportunism and strategic complacency on both sides ... the sheer scale of AUKUS puts it in a class of its own as an exemplar of bureaucratic incompetence."[40]

We will leave the technical and capability arguments to Professor White and others. But one point which we will highlight, based on our military experience over many decades, is that the transition complexity for completing a life-of-type-extension on the existing Collins Class boats whilst managing the introduction of around three used U.S. Virginia Class nuclear boats (with their reported maintenance / availability issues) whilst continuing to manage a problematical surface fleet ship construction program and concurrently introducing a yet to be designed UK Nuclear Boat by a Navy that is already 6% below authorised staff levels and is experiencing recruiting difficulties is a recipe for disaster.[41]

Given Australia's Defence procurement history, the possibility of this occurring on schedule and on budget is close to zero. However, this will surely provide excellent lessons to be ignored in a future Defence review!

THE IMPACT OF THE AUSTRALIAN POLITICAL SYSTEM ON NATIONAL SECURITY

Given our previous assessment of the Australian political system's impact on defence, we will now look at the impact on national security, a much bigger canvas for politicians to muddy.

Our first concern is that the analysis of our national security continues to be conducted through a somewhat narrow military platform lens.

For example, the commitment to what appears to a 25-year plus program to acquire two different types of nuclear submarines from two separate nations, for a cost in the order of $350B is being justified with little opportunity for the Australian people to understand why the decision was taken.

This is particularly problematic considering that Australia lacks a national risk assessment, a national security strategy, and a national defence strategy. The latter, however, was delivered in early 2024, three years after the AUKUS decision was announced.

All of this has transpired in a highly complicated security context, with uncertainty in global economic systems already generating palpitations in many countries. This is the result of a broken political system over the preceding decades, which repeatedly fails to learn from previous mistakes and failures.

This situation has been described by former Liberal Party opposition leader John Hewson as follows: "Voters are also increasingly tired of the adversarial nature of the contest between Liberal and Labor – pointless short-term point scoring unrelated to either the national interest or to addressing important issues. The main parties are too often focused on a power game played among themselves, saying, or doing whatever they believe is necessary to win, where the endgame has little to do with listening to voter concerns or providing better government."[42]

Our second concern is the oft venerated ANZUS Treaty; the negotiation of ANZUS in 1951 was the first instance of Australia forming a political alliance without the involvement of Britain, causing some tension with 'the motherland'.

Australia's commitment to U.S.-led causes such as the Vietnam War, while not formalised under ANZUS, have been linked to the treaty. Many political commentators suggested that Australia's strong involvement in Vietnam was a means of proving our usefulness to the alliance, demonstrating Australian loyalty should we need U.S. support in the future.

Despite being in operation for more than 60 years, the ANZUS treaty has only been formally invoked once in 2001 as a response to the September 11 terrorist attacks in New York and Washington, leading to Australia's long-term involvement in the United States led 'War on Terror'.

The ANZUS treaty has occasionally experienced difficulties.

Australia expressed disappointment at the minimal support received from the United States during the Konfrontasi conflict in Indonesia and Malaysia in the early 1960s.

There has also been ongoing discussion about the contemporary relevance of an alliance that was created in a 1950s Cold War environment. Since the early 1980s the United States has effectively absolved itself of any obligation to New Zealand when the latter became a nuclear-free country and refused entry to its ports by potentially nuclear-capable US warships.

Today, Australia maintains bilateral security relationships with both New Zealand and the United States under the treaty's banner. However, while the treaty has not been formally revoked after the breakdown of relations between the United States and New Zealand, the treaty between the three nations no longer fully exists in practice.[43]

Our third concern is the cost, past, present, and ongoing, of our political decisions. The dismal and costly results of Australia's involvement in the Vietnam, Iraq and Afghanistan wars is the result of repeated failures of political leaders of many nations and the inability to conceive of any alternative to address persistent security concerns.

The authors have worked closely with the U.S. armed forces and have respect and admiration for the men and women of the U.S. Services, but less for the parade of politicians from both the U.S. and Australia.

So, what was the cost of Australia's commitment to the U.S. Alliance?

- Vietnam - Australia suffered 521 deaths and a little over 3,000 wounded during the Vietnam War. 60,000 Australian soldiers were sent to the front, with a third of them being conscripts. The costs associated with participation were substantial, in the order of $220M, including both the direct costs of deployment and support for South Vietnam.[44]

- Iraq – Around 2000 ADF personnel served in this conflict. No Australian military personnel were killed in direct combat action during Operation Falconer or Operation Catalyst; however, one Australian soldier died in 2015 as a direct result of injuries sustained in an IED blast in 2004. The cost of the Iraq war to Australian taxpayers is estimated to have exceeded A$5 billion.[45]
- Afghanistan - the Afghanistan conflict resulted in 41 Australian military deaths, significant injuries, and a financial cost exceeding $10 billion; around 30,000 ADF personnel served over two decades in Afghanistan.[46]

What is harder to estimate is the ongoing physical and mental health impacts on the men and women who served in these wars. When compared to the wider Australian population, the age-adjusted rate of suicide was 26% higher for ex-serving males, and 107% higher for ex-serving females. The rate for ex-serving males who separate involuntarily for medical reasons is over three times higher than those who separated voluntarily.[47]

Even to this day, the Government, Defence, and the Department of Veterans' Affairs have been unable to address the breadth and depth of concerns in the veteran community.

This list also does not account for the millions of civilians who died as a result of these conflicts.

The loss of the Vietnam war, the eventual return of the Taliban to power in Afghanistan after a chaotic withdrawal of U.S.-led coalition forces, and the confused security situation following the Iraq wars leads us to ask the question: for what outcome did these Australian men and women sacrifice themselves, their ongoing health, and the wellbeing of their families?

For the hope of a 'guarantee of protection' from a great and powerful friend? A great and powerful friend whom we are now following into a potential conflict with China, under the AUKUS banner, without any public debate regarding our lack of a National Security Strategy to justify the action taken by the previous Morrison Government? A potential conflict for which there is little to no understanding in the Australian people of the impact of such a conflict on the world and our way of life.

The authors of this chapter are not saying this as "naïve anti-war advocates." We each have decades of service in the ADF. We say this as former military officers, conscious of the cost of conflict to the men and women in the ADF, and to the broader community, and as Australian citizens concerned with the short-term, reactive, political culture in our nation.

Finally, the question of the lack of accountability. Governments often measure progress by expenditure and not results. We need to measure national and defence capabilities by results against milestones. The liberal use of the terms 'soon and sooner' by our Defence portfolio Ministers are political tools, not milestones.[48]

THE NEED FOR A NATIONAL SECURITY STRATEGY

Our IIER-A National Resilience Project_posed three fundamental questions: What is a resilient society? Are we resilient enough? Can we make ourselves more resilient?[49]

We postulated three key attributes of a resilient society:

- **Shared Awareness / Goals.** With shared awareness we can act rationally and prepare accordingly because we can then define a shared goal - a common aimpoint; without it, we just react to each crisis as it occurs.
- **Teaming / Collaboration.** We cannot solve our complex challenges by looking for incremental, stove-piped, quick wins; we need a team approach within our nation and, as importantly, with our neighbours and allies.

- **Preparedness / Mobilisation.** There is no verb for 'resilience'; the verb 'prepare' is the most relevant in this case. As a nation we need to prepare for future disasters / crises and not just wait to react. In addition to preparing, we must be able to mobilise the nation to address an emerging threat.

We fail in our approach to national security at the first step, that is, of shared awareness. Increasing levels of misinformation in both professional and social media, combined with extensive use of political spin have resulted in higher levels of complacency, confusion, and disengagement in our society.

Political divisiveness further diminishes the ability to build shared awareness and shared goals in our nation at a time where the need to do so has never been greater.

Our assumptions frame our resilience and therefore they also need to be made explicit.

Short-termism, or 'quick win' thinking, is deeply embedded in the Australian political culture; collectively we tend to focus on today and largely on our personal needs, not on future interacting and cascading risks that will impact our whole society.

Thirty years of relative prosperity in Australia, fuelled by lower trade barriers, privatisation, and deregulation, have increased our productivity and wealth, providing the resources necessary to address the challenges we confront today if we choose to act.

However, many of these challenges are themselves a result of globalisation, e.g. our extensive reliance on overseas supply chains for critical goods, leaving our nation vulnerable. Whilst the lower cost of goods has had economic and standard of living benefits, there is a very high price to 'cheap' in a crisis.

Australian Governments have readily accepted responsibility for national security; however, since 2013 no federal Government has had a national security and resilience strategy.

This is an unacceptable situation in today's complex world,

disrupted by the pandemic and facing growing global economic and security challenges.

The DSR 2023 is understandably focussed on our regional security concerns. However, we should also pay heed to the comments of European leaders such as Polish Prime Minister Donald Tusk, who said Europe is entering a 'prewar' era, cautioning that the continent is not ready and urging European countries to step up defence investment ..."I don't want to scare anyone, but war is no longer a concept from the past. It's real and it started over two years ago."[50]

Conflict in Europe, the Middle East and in the Asia Pacific region seems as unthinkable today as it was in the late 1930s. But it happened, and the same warning signs are here today.

The failure to grasp the need for a comprehensive national risk assessment to underpin National Security Strategy leads our politicians to default to platform level solutions to a security crisis.

Announcing the acquisition of a small number of nuclear-powered submarines, that may not achieve full operational capability for another 20 years, will do little to address the security challenges we face today and into the near future.

Whilst this is a grim assessment there is some cause for cautious optimism. The actions we need to take are not beyond our ability to design and implement.

We have considerable expertise, will and resources in this country. In these times of uncertainty, we, the Australian people, need to act and demand more of our political system.

We should start with a comprehensive national risk assessment and a National Security Strategy, and we should hold our politicians accountable for delivering those in practicable, and not simply political or rhetorical, ways.

APPENDIX 4: ADDITIONAL ANALYTICAL ASSESSMENTS

THE AUSTRALIAN SUBMARINE DECISION: A LOOK BACK

October 3, 2024

By Pierre Tran

Paris – It is all about governance – or the lack thereof – might be one way to summarise *Nuked: The Submarine Fiasco that Sank Australia's Sovereignty* (Melbourne University Publishing), a book which takes a critical look at Canberra's cancellation of a multi-billion dollar deal for French attack boats.

The author, former Australian Broadcasting Corporation reporter Andrew Fowler, presented the recently published *Nuked* Sept. 24 to the Paris branch of the National Union of Journalists (NUJ). French media and reporters of the Association des Journalistes de Defence press club attended as guests.

The "fiasco" refers to Australia, then led by prime minister Scott Morrison, announcing September 16, 2021 the axing of a deal with a French shipbuilder, Naval Group, to build 12 diesel-electric submarines for the Australian navy.

Morrison switched to the U.K. and U.S. as partners to deliver nuclear-powered submarines in the wider AUKUS project, dumping France and reversing a pick of diesel propulsion.

Nuked sets out an account of the political reasons for that policy switch, which the author said sought to tie Australia closer to America and develop nuclear power at home.

The significance of international submarine deals for Paris could be seen with the Netherlands signing Sept. 30 a contract reported to be worth €5.65 billion ($6.25 billion) with Naval Group (NG), with the French company to build four Orka class boats, replacing the four-strong Walrus fleet.

NG said in a Sept. 30 statement on the signature of the delivery agreement the "expeditionary" boats would strengthen "strategic capabilities" of the Royal Netherlands Navy. This was widely taken to mean fitting out the ocean-going Orka for the U.S.-built Tomahawk cruise missile, giving the Netherlands for the first time a long-range submarine capability to strike land-based targets, as well as torpedoes to hit ships.

NG pitched its Barracuda diesel-electric boat against rival German and Swedish bids, respectively ThyssenKrupp Marine Systems and Saab Kockums. The Netherlands authorities rejected an appeal lodged by TKMS, which contested the pick of NG.

The French shipbuilder signed Sept. 10 an industrial cooperation agreement, pledging to inject close to a reported €1 billion in transferring advanced submarine technology and expertise to companies in the Netherlands over 20 years.

"This contract will allow the Netherlands to deploy ocean-going submarines of world class standard, strengthening the armed forces of the Netherlands as well as the European capabilities in Nato," the French armed forces ministry said in an Oct. 2 statement.

Australian Party Politics

Back in Australia, there was a "failure of due process" on cancelling the planned French submarines and opting to work with the U.K. and U.S., Fowler said.

"I was appalled, not so much by the process of military acquisitions, which is usually a pretty murky, dirty, and dangerous world, but because what we experienced in Australia was an example of what we call 'executive overreach,'" he said.

The underlying issue was the lack of public and parliamentary debate on a critical decision, namely dropping a major arms deal, in which France had beaten rival German and Japanese offers "by a country mile," he said.

"Executive government decides to do something with public money and hides it from parliament, the process, and does what it wants to do for its own political purpose," he said. That switch to AUKUS meant Australia took on spending of A$368 billion in an "unaccountable venture," he said. Australian officials later told an inquiry they did not know why there was a requirement for a nuclear boat and how it would be financed, he said.

There was Australian party politics at play, he said, with the conservative Liberal Party, then in power, keen to strengthen ties with the U.S., which sought "containment" of China.

A pursuit of closer links with the U.S. served as a differentiator – or "wedge" – in national security policy for Morrison, looking to distinguish the Liberal Party from the opposition Labor Party, he said. Labour was anti-nuclear and took a more independent approach – similar to France.

The present center-left Labor administration has continued to pursue the AUKUS project. Australia is due to hold a federal election in 2025.

Australia-America

The other key factor which pointed up difference between the two parties was Liberal Party support of nuclear power, seen as a big break in Australian policy, he said.

Morrison's pursuit of deeper Australian-American cooperation pointed up a move from "cohabitation" to "interchangeability," with a view of "seamless joining" of forces, he said. Australian procurement of American atomic-powered boats offered two desired outcomes in one fell swoop, with France discarded by the wayside.

That tightening of ties made Australia dependent on U.S. policy, he said, at a time when tension had increased between Washington and Beijing.

The cancellation of an ocean-going fleet of conventional Shortfin Barracuda boats showed a big policy difference between France and Australia, he said, with Paris seeing the submarine as a strategic partnership, while Canberra saw it as a commercial deal.

An irony was the planned Australian Shortfin was a conversion of the French nuclear-powered Barracuda boat to diesel power, he said. That begged the question, if Australia wanted nuclear boats, why not go with France, which some said built the best nuclear submarines, powered by low enriched uranium?

That was because it was not just about nuclear submarines, he said, but supporting the U.S. in containing China. That could be seen as raising the risk for Australia, he added.

"Don't Mess Around"

Australia retained a British consultant, David Gould, in the competition for the initial submarine deal, he said. Gould pointed up the strength of the French bid – "Don't mess around, France has got it," Fowler said.

There was the German bid, which offered free access to the intellectual property, he said, while the French were very careful with control of advanced technology, which included "a secret propeller and very quiet boats."

French technology was only available "on licence," requiring Australia to work closely with French industry, he said. That was in contrast to the German offer, which offered full access to the IP.

"Has it occurred to you that the French IP is really worth a lot of money," Gould told the Australians, he said. "It's a closely guarded secret and they are sharing it with you...and the German IP isn't worth a pinch of salt. That's why they're giving it to you."

With the Liberal Party in power, Morrison took up the prime

minister's post in 2018, while his predecessor, centrist Malcolm Turnbull, was removed, or "rolled," from the top job.

Morrison was a hard rightwing, Christian Evangelical, Fowler said, and a close friend of Mike Pompeo, also a Christian Evangelical. Morrison took a hard line with the French, he said, and it was around that time the Australian press gave hostile coverage against the French-led Barracuda project.

"They were running a campaign against the French," he said. The Australian media reported the project was late and over budget, when a few months and the amount of money were insignificant in view of the scale of the program.

Such was the hostile climate, the Naval Group executive chairman, Pierre Eric Pommellet, flew 24 hours to Australia, stayed two weeks in Covid quarantine, and sought to meet Morrison, he said. That meeting did not happen, and Pommellet returned empty handed.

The NG chief executive offered 60 pct of spending in Australia on the boats, even though that was not in the contract, he said.

"It wasn't enough," he said.

Some eight months after that failed flight, the Australian office of national intelligence started secret talks with the U.S. state department and Pentagon on acquiring nuclear boats, he said. The Australians and Americans kept French officials in the dark on those critical discussions. It was only a matter of time before the Canberra ax fell, and it duly fell on Sept. 16, 2021.

Geostrategic Region

France sees the Indo-Pacific as a strategic region, with some 1.7 million French nationals living in overseas departments, territories, and other nations in that vast area.

The Indo-Pacific accounts for more than a third of French trade in goods outside the European Union, and that has grown by 49 percent in 10 years, the foreign ministry said on its 2021-2022 strategy for the region.

The French services have deployed 8,300 personnel as pre-positioned forces.

The perceived importance of the region can be seen in the annual Pégase exercise flown by the French air force to the Indo-Pacific. The service completed in August its 2024 exercise. French aircrews flew with the Royal Air Force in the Griffin Strike exercise to Australia for the first time, as part of the Pégase mission. The French service flew three Rafale fighter jets, two A330 Phénix multirole tanker transports, and two A400M Atlas military transports, while the RAF flew six Eurofighter Typhoons, one A330 MRTT and one A400M Atlas.

The British and French aircrews flew some 15,000 km to Darwin air base, taking four days and two stop overs, and took part in the Australian Pitch Black 2024 exercise along with other allied nations.

French aircrews also flew with German and Spanish pilots in another part of the Pégase exercise.

These three allies are partners in the future combat air system (FCAS) project, which comprises a new generation fighter, remote carrier drones, and a combat cloud communications network.

China Ever Present

The perceived importance of China can be seen by how often that nation is on the front page.

On the military front, there was an unusual Sept. 25 test firing of an intercontinental ballistic missile, with dummy warhead, into international waters. Australia, New Zealand, and Japan said China had failed to give advance notice of that launch, which caused grave concern.

On the same day, Reuters news agency ran its exclusive story that a Russian company, IEMZ Kupol, a subsidiary of Russian state-owned arms company Almaz-Antey, was developing and building a long-range attack drone in China, with local help. That weapon was due to be used in Ukraine.

The French stock market closed Sept. 26 sharply higher, with the CAC-40 index up 2.3 percent. That unusual gain was reportedly in response to China's measures to inject life into its fading economy, prompting equity traders to snap up stocks in French luxury brands.

On the political front, the BBC reported Sept. 26 Donald Trump, the Republican candidate in the presidential election, had referred to China 40 times in five rallies in the presidential debate last month. Trump mentioned China 27 times in a town hall meeting in Michigan in the previous week, the broadcaster said.

Those references tended to highlight "tension" between the U.S. and China, with the latter seen as a "kind of economic predator," the report said.

ROYAL AUSTRALIAN NAVY EXPLORES AUTONOMY AND OPTIONAL CREWING: EYES LUSV AS POTENTIAL LOSV SOLUTION

May 12, 2024

By Gregor Ferguson

The Australian Department of Defence's response to the Royal Australian Navy surface fleet review, *Enhanced Lethality Surface Combatant Fleet*, published in February, announced the RAN would field six Large Optionally Manned Surface Vessels (LOSVs) from the 2030s to carry missile launch systems.

Interestingly, the RAN program closely resembles a similar but far more advanced one in the United States, the U.S. Navy's Large Unmanned Surface Vessel (LUSV) program.

The LOSVs will each carry a 32-cell Mk41 Vertical Launch System (VLS) to supplement the firepower of the RAN's manned warships. Its Lockheed Martin Aegis Baseline 9 combat system makes possible a Cooperative Engagement Capability (CEC) with Aegis-equipped manned surface ships and enables the vessel to carry out both Anti-Ballistic Missile (ABM) and conventional Anti-Air Warfare (AAW) operations.

If the LOSV program goes ahead, the ships will augment the RAN's three Hobart-class AAW destroyers and six Hunter-class Anti-

Submarine Warfare (ASW) frigates – which will still have a world-class AAW sensor and combat system combination – as well as the seven to eleven General-Purpose frigates to be acquired from later this year. None of these ships are over-endowed with Mk41 VLS launch cells: the Hobart-class ships have 48, the Hunter-class has 32 and the General-Purpose frigates will have just 16.

Importantly, the LOSV will be equipped with the Aegis Baseline 9 combat system: Aegis alone comes with the coveted CEC which the U.S. Navy has only ever shared with Japan and Australia. All of the RAN's Aegis-equipped warships are also being upgraded to Baseline 9 which enables ABM defence and protection of a sea or contiguous land area against hypersonic ballistic weapons.

So the LOSVs could operate in direct support of these ships – but they could also protect deployed Australian troops, Australian and allied ships threatened by hypersonic ballistic anti-ship missiles or even vulnerable Australian population centres.

The LOSVs obviously need to have an Aegis-equipped ship within communications range with a sensor suite that can detect conventional missiles, aircraft and ballistic missiles. The SPY-1D(V) radars on the Hobart-class can do this, and so can the CEAFAR2 radars on the Hunter-class.

We don't know yet about the General-Purpose frigates, though most builders of the contenders at last year's Indo pacific 2023 show in Sydney showed models of their ships with the CEAFAR2 radar and stated they used the Saab Australia 9LV Mk3E combat system/tactical interface which is used universally by the RAN.

The Mk41 VLS will enable anything from 128 quad-packed Evolved Sea Sparrow Missiles (ESSM) for self-defence, to 32 single-packed SM-2, SM-3, SM-6 or Tomahawk missiles for long-range anti-aircraft, anti-missile/ABM and strike purposes. As long ago as 2021 the U.S. Navy conducted a successful CEC-enabled trial of Raytheon's AMRAAM-based SM-6 aboard its Unmanned Surface Vessel (USV) Ranger using a Mk41 VLS as the launcher.

While an Australian-built autonomous LOSV may seem like a

distant dream for most Australians, the reality is actually much firmer.

The U.S. Navy and RAN have both proven the CEC works using the Aegis Baseline 8 system. Just as important, fully autonomous and optionally crewed vessels have been tested by both navies: while the U.S. Navy has had prototypes in the water for several years, Perth-based Austal Australia put an optionally crewed prototype to sea for the first time in March this year and finished the Sea Acceptance Trials (SAT) on this vessel in April.

The successful SATs (including Endurance Trials) of the remote and autonomously operated vessel *Sentinel* marked the first phase of what the RAN calls the Patrol Boat Autonomy Trial (PBAT). They consisted of a series of remote and autonomous navigation events conducted off the Western Australian coast during March and April 2024.

The trial vessel, the RAN's de-commissioned Armidale-class Patrol Boat (ACPB) HMAS *Maitland*, employed Perth-based start-up Greenroom Robotics' Advanced Maritime Autonomy Software to navigate reliably. At 57 metres LOA, Sentinel is by far the largest vessel operated in Australia to be operated remotely and autonomously.

Funded (at an undisclosed level) by the Commonwealth of Australia, PBAT is a collaboration between Austal Australia, Greenroom Robotics, the Brisbane-based Trusted Autonomous Systems Defence Cooperative Research Centre (TAS DCRC, which receives its core funding from the Department of Defence) and the Royal Australian Navy's Warfare Innovation Navy (WIN) Branch. Its aim is to use the former ACPB to provide a proof-of-concept demonstrator for optionally crewed or autonomous operations.

The release in 2020 of the RAN's Robotic and Autonomous Systems – Artificial Intelligence (RAS-AI) 2040 Strategy was the trigger for the TAS DCRC and Austal to get together. They saw an opportunity to re-purpose the former HMAS *Maitland* to define and better understand existing autonomous technology and how it could meet RAN needs. The partners also saw an opportunity to explore whether an autonomous platform could deliver asymmetric warfighting advantage.

Austal took possession of the decommissioned HMAS *Maitland* in 2022 and modifications included changes to the ship's navigation, communications, bilges, CCTV, and electrical systems.

The *Sentinel* has been fitted with two autonomy systems: firstly, GreenRoom Robotics' software-based GreenRoom Advanced Maritime Autonomy (GAMA) system, which enables remotely operated or full autonomous missions while complying with the International Regulations for Preventing Collisions at Sea, COLREGs, without crew intervention.

Secondly, a platform autonomy system, developed by Austal and based on the company's in-house MARINELINK control and monitoring system, which enables operation of the *Sentinel's* mechanical and electrical systems without crew intervention.

A key aspect of the initial trial was the endurance component, designed to observe *Sentinel's* behaviour in an extended endurance mode. During this trial she operated autonomously with minimal to no crew interaction.

When the ship returns to sea the PBAT trial will focus on a bunch of other goals, starting with simply progressing the concept of remote operations and the autonomous certification approach.

At a sub-system level, the partners need to investigate and understand the sustained operation of shipboard mechanical systems reliably without crew intervention, including adding redundancy to enable operations at sea for extended periods, something the U.S. Navy has explored also. They also need to understand better how fuel management, communication and navigation systems can be made autonomous, and how they will work.

Longer-term, the PBAT trial will generate data contributing to risk reduction for future RAN projects involving remote or autonomous vessels. Short-term, the RAN could potentially transfer lessons learned about remote and autonomous systems to its current fleet to optimise crew workload: remote and autonomous operation has the potential to reduce crew workload and increase operational safety by reducing human error.

Austal says it is open to expanding the PBAT program and actively investigating opportunities to both extend current autonomy and optional crewing systems and integrate new systems to increase *Sentinel's* capability or that of any future trials vessel.

Any future phases will be assessed to ensure the needs of the RAN's Robotics and Autonomous Systems-Artificial Intelligence (RAS-AI) 2040 Strategy are still being met, the company says. Extension of this program will help to build the capabilities necessary to support two key requirements, says the company: the future LOSV program (which is where we came in); and the introduction of new technologies into the broader surface fleet to ensure future crewing requirements can be achieved. These aren't stated as yet but the firm trend in Australia is to use fewer personnel and have smaller ship's companies.

The PBAT program wasn't established specifically with the LOSV program in mind, points out Austal, though everything to do with LOSV will benefit from the PBAT trial. Austal strongly supports the introduction of the LOSV to the Navy's surface fleet, as you'd expect given that it will build the ships; and it points out also that its investment in autonomy, both in Australia and in the USA, has been with this type of platform in mind.

Austal's Chief Executive Officer Paddy Gregg said the completion of the initial phase of the sea trials marks a significant PBAT milestone, successfully demonstrating the capability of the locally developed autonomous systems and their integration within a full-size, Australian-made naval vessel.

"Looking ahead, we are excited about the potential opportunities to work with [the RAN] to further advance the autonomous technology demonstrated during the trial; on projects such as the Large Optionally Crewed Surface Vessels (LOSV), recently announced by the Australian Government as part of the Surface Combatant Fleet Review," Mr Gregg said in a statement.

So, this work positions the RAN to adopt autonomous technology in the future. In recent announcements Australian Defence Minister Richard Marles has said the planned LOSV platform will likely be acquired through formal RAN engagement with the U.S. Navy's LUSV program. Essentially, whatever the U.S. Navy gets Australia will get, he's suggesting, and the RAN will be a 'fast follower'.

But however enticing the LUSV program looks, the PBAT trial is also designed to address Australia's own sovereign requirement for a trusted autonomous system, especially an armed one. It needs to ensure the autonomous control system aboard the LOSV conforms with Australia's needs and with the country's high ethical standards for robotic and autonomous systems.

LUSV Program

The U.S. Navy's Large Unmanned Surface Vessel (LUSV) program is, as you'd expect, very similar to the RAN's LOSV program. It is designed to deliver adjunct missile magazine capacity – essentially Mk41 VLS cells – to the Fleet as part of the U.S. Navy's Distributed Maritime Operations (DMO) concept. The difference is that the U.S. Navy has been experimenting and developing technology in this area for several years and (budgets permitting) plans to order its first production autonomous ship as early as FY2025 at a planned cost of about US$315 million.

The U.S. Navy's vision for LUSV is for a ship between 200 and 300ft LOA with a full-load displacement of approximately 1,500 tons. It is intended to be a low-cost, high endurance, modular USV that can carry a variety of payloads.

Late last year U.S. Naval Sea Systems Command issued a Request for Information (RFI), asking industry for feedback on its draft LUSV proposals. "[It] will be built to commercial American Bureau of Shipping (ABS) vice military standards," said the RFI which closed in December. "As an adjunct magazine, LUSV will operate with Carrier Strike Groups (CSG), Expeditionary Strike

Groups (ESG), Surface Action Groups (SAG), and individual manned combatants."

"The LUSV will be capable of autonomous navigation, transit planning, and COLREGS compliant maneuvering and will be designed with automated propulsion, electrical generation, and support systems," according to the U.S. Navy's FY 2024 budget documents.

"LUSV missions will be conducted with operators in-the-loop (with continuous or near-continuous observation or control) or on-the-loop (autonomous operation that prompts operator action/intervention from sensory input or autonomous behaviors)."

The Navy plans to issue its formal requirement for a production LUSV this calendar year. The PEO Unmanned and Small Combatants and PMS 406 are leading the U.S. Navy's effort.

The LUSV program started to gain traction in 2020 when the service awarded LUSV study contracts worth US$42 million to six US companies: Austal USA; Bollinger Shipyards; Fincantieri Marinette; Gibbs & Cox; Huntington Ingalls Industries (HII); and Lockheed Martin.

In March this year the U.S. Navy announced that propulsion plants for solutions expected to be offered by four of the six – Bollinger, Fincantieri Marinette, HII and Gibbs & Cox – had all passed the mandated 720-hour engine reliability tests. These were intended to demonstrate that different propulsion plants can operate for extended periods without human intervention. This test is the milestone the LUSV program must pass before it can go into a formal development phase.

Meanwhile, autonomous ships already operated by the Navy have surpassed or come close to the 720-hour benchmark. A fleet of four unmanned prototypes – Unmanned Surface Vessels (USV) *Mariner*, *Ranger*, *Seahawk* and *Sea Hunter* – from the U.S. Navy's San Diego-based Unmanned Surface Vessel Division One (USVDIV-1) self-deployed across the Pacific to Sydney late last year to participate in the Australian Defence Force's (ADF) EX Autonomous Warrior in

October-November 2023. Later, they headed on to Japan before returning home.

The role of Austal USA is interesting. It is based in Mobile, AL, but is a subsidiary of Austal Australia. Last year the parent company signed the Heads of Agreement with the Australian government which could result in Austal becoming the government's strategic shipbuilder at its Henderson base near Perth under a new Strategic Shipbuilding Agreement.

So, any ships built in Western Australia for the RAN would be built by Austal, including the LOSVs and between four and eight General-Purpose Frigates announced in this year's Australian Surface Combatant Review.

Austal has been positioning itself carefully for the new world of autonomy. A year ago the company handed over to the U.S. Navy an optionally crewed Expeditionary Fast Transport, USNS *Apalachicola* (EPF-13), the largest ship in the U.S. fleet with autonomous capability. And in January, with L3Harris, it launched the OUSV-3 *Vanguard*, the first autonomous ship designed for the U.S. Navy from the keel up, though sister to two existing USVDIV 1 ships, *Ranger* and *Mariner*.

There are no suggestions that Austal's participation in the RAN's PBAT trial will have any effect on its likely bid to build the LUSV for the U.S. Navy, but its investment in research across two nations in autonomy and robotics won't do it any harm at all.

AUSTRALIAN DEFENCE FORCE EMBRACES LONG-RANGE FIRES

June 23, 2024

By Gregor Ferguson

It's a fact: in any conflict the fighter with the longer reach has a definite advantage. The Australian Defence Force is trying to create that advantage for itself by spending up to $27 billion on 12 new families of long-range missiles between now and 2034 and investing another $21 billion in manufacturing at least some of them in-country.

And it plans to spend another $7.3 billion developing what April's Integrated Investment Program (IIP) calls a Defence Targeting Enterprise to detect, identify, track and allocate weapons to those targets, engage them and then carry out post-strike reconnaissance. Essentially, the ADF is looking to go beyond the rhetorical limitations of the 'Kill Chain' and establish a 'Kill Web' which enables it to address a much wider area than hitherto. More about this below.

Some of those missiles and targeting systems are coming together this calendar year: the first of four Northrop Grumman MQ-4C Triton High Altitude Long Endurance (HALE) Uncrewed Air System (UAS) ordered by the RAAF has just arrived at RAAF Base Tindal in the Northern Territory. The first of the Kongsberg Naval Strike Missiles, which will replace the venerable Boeing AGM-84-series Harpoon that arms the RAN's Hobart-class destroyers and ANZAC-class frigates is also due to arrive in 2024. And the first Australian-made GMLRS rockets will be flight tested in 2025, with a batch of US-made GMLRS and their HiMARS launchers entering service at around the same time.

The implications of this dramatic extension in strike range are significant. If it gets this right, the ADF will be able to hold an aggressor at risk at some distance from its home soil and its own forces; but it also needs to refresh both its targeting systems and its diplomacy to reflect its much greater reach and impact. This is all consistent with the ADF's transition from a balanced force to an integrated force and a new defence strategy of Deterrence through Denial.

However, those long-range weapons aren't in service with the ADF yet. At present, Army's reach is about 30-40km using conventional artillery; the RAN's reach is 170km using the SM-2 anti-air and -missile weapon, or about 125km using the anti-ship Harpoon Block II.

Table 1 lists the long-range weapons currently in service with and sought by the ADF. Except for the Lockheed Martin GMLRS and some naval anti-air/missile weapons, the minimum range of the new generation of missiles is 200km.

TABLE 1 – Long Range Weapons in service with or sought by Defence

Missile	Manufacturer	Estimated Range	Application	Launch Platform
RGM-109E Tomahawk	Raytheon	1,600km +	Long-range strike, anti-ship	Hobart, Hunter and Virginia-class ships and submarines
Naval Strike Missile (NSM)	Kongsberg	250km+	Anti-ship, strike	Hobart, Hunter, ANZAC and General Purpose frigate
Joint Strike Missile	Kongsberg	500km+	Anti-ship, strike	F-35A Lightning II
GMLRS	Lockheed Martin	70km+	Tactical strike	HIMARS
Precision Strike Missile (PrSM) – Increment 2	Lockheed Martin	500km+	Long-range strike, anti-ship	HIMARS
Precision Strike Missile (PrSM) Increment 4	Lockheed Martin/Raytheon-Northrop Grumman	1,000km+	Long-range strike	To be determined
AGM-158A Joint Air-Surface Standoff Missile (JASSM)	Lockheed Martin	370km+	Strike	F/A-18F Super Hornet, F-35A Lightning II
AGM-158B Joint Air-Surface Standoff Missile (JAASM-ER)	Lockheed Martin	800km+	Long-range strike	F/A-18F Super Hornet, F-35A
AGM-158C Long-Range Anti-Ship Missile (LRASM)	Lockheed Martin	370km+	Anti-ship	F/A-18F Super Hornet, F-35A Lightning II, P-8A Poseidon
AGM-88G Advanced Anti-Radiation Guided Missile (AARGM-ER)	Raytheon	300km+	Anti-radar, anti-communications	EA-18G Growler and F-35A
Hypersonic weapons	various	To be determined	Strike, anti-ship	HIMARS, F/A-18F Super Hornet
OWL (One-Way Loitering munition)	Innovaero	200km+	Strike, anti-ship	To be determined
Land-based anti-ship missile - NSM/PrSM Increment 2	Kongsberg-Thales or Lockheed Martin	250-500km	Anti-ship; strike also in the case of PrSM	Bushmaster PMV, HIMARS
RIM-162 Evolved Sea Sparrow Missile (ESSM)	Raytheon	50km+	Naval self-defence	Hobart, Hunter, ANZAC, General Purpose frigates
RIM-66 Standard (SM-2)	Raytheon	170km+	Long-range naval anti-air and -missile	Hobart, Hunter, LOSV
RIM-174 Standard ERAM (SM-6)	Raytheon	370km+	Long-range Anti-Ballistic Missile (ABM) and anti-air, -missile and -ship defence	Hobart, Hunter, LOSV

Only the RAAF currently has a genuine long-range strike weapon, Lockheed Martin's AGM-158A Joint Air-Surface Stand-off Missile (JASSM) which arms the service's F/A-18F Super Hornets and P-8A Poseidon patrol aircraft and has a range of more than 370km. This stealthy weapon entered service in 2006 (it will be supplemented, and then replaced, by the AGM-158B Extended Range version of JASSM which has an 800km range) and has given the ADF some valuable experience of targeting a weapon with such a range.

The RAN and RAAF launch platforms and their general organisation won't change when these services field the new weapons. But the Army has begun a significant transformation to operate this new generation of stand-off missiles. The long-established 16th Regiment, Royal Australian Artillery (RAA), operates the Saab RBS-70 point defence missile; from next calendar year this will be replaced by the eNASAMS (enhanced National Advanced Surface-Air Missile System) acquired under Project LAND17 Ph.7b. Primed by RTX subsidiary Raytheon Australia, this will blend the proven Raytheon AIM-120 AMRAAM and AIM-9X missiles and Kongsberg launchers with the CEATAC and CEAOPS phased array radars manufactured by Australian company CEA Technologies.

The 16th Regiment will be part of an all-new formation, the Adelaide-based 10th Brigade which was formed in January 2024 and whose focus is on long-range fires and air and missile defence. From 2025 the Brigade's new 14 Regiment will operate 36 HiMARS launchers, acquired from Lockheed Martin under a US Foreign Military sales (FMS) agreement, firing the GMLRS and its extended range variant. The remaining six launchers will be used for training. The capabilities of the Brigade's planned 15 Regiment, its launchers and projectiles, are still under Defence consideration; they will be delivered under Project LAND8113 Ph.2.

In due course (Defence hasn't said when) 14 Regiment will also operate Increment 2 of Lockheed Martin's Precision Strike Missile (PrSM) which has a range of 500km. And Australia is a partner in developing the PrSM Increment 4 missile which is still under development (Lockheed Martin is battling for this contract with a Raytheon-Northrop Grumman consortium) and will have a range of 1,000km.

The Australian Army also has a requirement for 30 land-based anti-ship missile systems to be operated by the 10th Brigade, originally under Project LAND4100 Ph.2 – Land-Based Maritime Strike. However, the Brigade's first land-based maritime strike capability will be delivered by 14 Regiment's HiMARS launchers, GMLRS and PrSMs under Project LAND8113. Phase 2 of this project will also see the Brigade's new 9 Regiment operate an undisclosed number of advanced radars – probably CEA Technologies' CEATAC and CEAOPS sensors.

While PrSM may eventually be able to hit moving targets such as ships, Kongsberg and Thales have spotted an opportunity and combined to offer the NSM-based Strikemaster anti-ship and strike system. This would see the missile fired from the same 'twinpack' launcher used in Kongsberg's proven Coastal Defence System and mounted on the 'pickup' version of Thales's Bushmaster wheeled armoured vehicle. Except for the Bushmaster, this is also very similar to the US Marines' autonomous, uncrewed Ground-Based Anti-Ship Missile system which uses the NSM with the same twinpack launcher.

The NSM has a 250km range from a ground or maritime launcher and will arm the RAN's three Hobart-class destroyers and six remaining ANZAC-class frigates, its six future Hunter-class frigates and possibly also its planned fleet of up to 11 General Purpose frigates. So, the Strikemaster looks a low-risk option for the ADF.

The RAN's long-range weapons will be the NSM, SM-2 and SM-6, and the RGM-109E Tomahawk strike missile. The latter has a range of more than 1,600km while the SM-6 has a range of around 370km.

The RAAF's air-launched strike inventory started with JASSM; its reach will grow with Lockheed Martin's JASSM-ER (800km range) and the AGM-158C Long-Range Anti-Ship Missile (LRASM – 370km range);

the Northrop Grumman AGM-88G Advanced Anti-Radiation Guided Missile (AARGM – 300km range); and possibly the Kongsberg Joint Strike Missile (JSM), the air-launched variant of NSM, which has a 550km+ range and has just been ordered by the USAF for its own F-35As. The JSM can be carried internally to preserve platform stealth. (In each case, range figures depend on the flight profile followed by the weapon).

These weapons will be carried by the RAAF's combat aircraft, the Super Hornet, EA-18G Growler, Lightning II and P-8A Poseidon patrol aircraft.

There's another contender also: Northrop Grumman is partnered with Raytheon Technologies to develop the Hypersonic Attack Cruise Missile (HACM) for the U.S. Air Force (USAF). HACM is reportedly a first-of-its-kind weapon developed in conjunction with the Southern Cross Integrated Flight Research Experiment (SCIFiRE), a U.S.-Australia project arrangement. Raytheon Technologies and Northrop Grumman have been working together since 2019 to develop and integrate Northrop Grumman's scramjet engines onto Raytheon's air-breathing hypersonic weapons.

Defence Targeting Enterprise

Targeting for these weapons is the ADF's new challenge and one of the keys to effective deterrence..

Defence needs to be able to detect, identify and track targets for its weapons; it may then need to indicate the target to a missile; and after the attack it needs to carry out post-strike reconnaissance to establish whether the target was actually hit and, if so, how badly it was damaged.

That bald statement of targeting principles barely describes the complexity of the targeting process. A sensor that detects and tracks a target isn't enough: they all need accurate positional data and, in the case of weapons with a moving target capability, information about the speed and direction of travel of the target. And the ADF needs corroborating intelligence and continuous situational awareness in order to determine what *not* to shoot at: the reputational damage, not to mention the financial and opportunity cost, of striking the wrong target may be massive.

This is what the IIP had to say in Chapter 5 about its planned Defence targeting enterprise:

5.2 An advanced and resilient network of sensors and communications and intelligence systems will be brought together to form a Defence targeting enterprise. The Defence targeting enterprise will provide Defence with the timely ability to detect, identify and track targets more precisely and at longer ranges in highly contested operating environments. The Defence targeting enterprise will be underpinned by a highly trained workforce and will be interoperable with the capabilities of the United States and other key partners.

The IIP allocates up to $7.6 billion to this program.

Defense.info asked Defence for further information about the Defence Targeting Enterprise, in particular whether or not a new HQ would be established within the ADF just to handle targeting. The response was unequivocal: the Defence Targeting Enterprise will not be a separate command within the ADF or wider Department of Defence. "Targeting," we were told, "will remain a process that is undertaken by commanders at different levels, enabled by the effective integration and augmentation of supporting capabilities." In response to *Defense.info*'s questions a Defence spokesperson said:

Targeting is an activity that has long been undertaken by Defence at different operational levels and in a variety of settings. Many organisations and systems contribute to it. The direction to establish a Defence Targeting Enterprise reflects a focused effort to modernise and enhance capabilities that enable the targeting process, ensuring they are optimised to support new long-range weapons entering service and the demands of an increasingly complex operating environment.

The Defence Targeting Enterprise is not a specific organisation that will be established in a particular place. Rather, it will be a federation of integrated organisations, systems and processes across Defence that, collectively, will allow selection and prioritisation of targets and match them with appropriate responses. This approach recognises growing interdepen-

dencies between different elements of Defence during the targeting process and will allow greater assurance of system efficacy. Many elements that will contribute to the Defence Targeting Enterprise already exist and will be modernised or enhanced. Other elements will need to be created or adjusted in coming years. This will not occur in a singular location and will evolve over time.[1]

In a separate email, a Defence spokesperson added:

Detection capabilities and the ability to sense underpin strike and IAMD capabilities. To enable the integrated, focused force to conduct enhanced long-range strike, Army will draw from an advanced and resilient network of sensors and communications and intelligence systems through the Defence Targeting Enterprise, including multi-mission phased array radars. 9 Regiment [of 10 Brigade] will operate advanced radars that will be acquired through the second phase of Land 8113 to provide an organic sense capability. This will be complemented by 10th Brigade's technical integration as a component of the Defence Targeting Enterprise, capable of providing and receiving target details from the broader force. [2]

As stated earlier, the ADF has been developing a stand-off targeting capability since it selected JASSM in 2006. Details are scarce, but the ADF's targeting resources will eventually include four (and possibly seven) Triton UASs with radar and infrared sensors; space surveillance sensors; the Jindalee Operational radar Network (JORN); radars and infrared sensors on the P-8A Poseidon and E-7A Wedgetail; MC-55A Peregrine ISREW aircraft; warships and their helicopters; Super Hornets and Growlers; F-35A Lightning IIs and all sensors operated by 10th Brigade. These all contribute to Situational Awareness but are also target detectors and trackers. The power to actually prosecute a target, however, will still reside in the hands of men and women – aloft, afloat or in a headquarters or command post somewhere.

The RAAF's first Triton is in-country. The RAAF has ordered four but really needs all seven it says it wants in order to field a robust capability. From an altitude of 55,000ft Triton has a radar horizon of more than 500km so its Over The Horizon (OTH) targeting capabilities are immense. Its mission endurance of as much as 30 hours also enables it to surveil up to 4 million square nautical miles of sea or littoral in a single mission. While the P-8 and the RAN's MH-60R Seahawk helicopters can also do surveillance and OTH targeting, they are limited in both time and operating area and can be re-tasked as contingencies emerge or change – Triton offers genuine persistence.

However, all three will feed into the Defence Targeting System: a naval contact spotted by a Seahawk, for example, could become a target for a land-based anti-ship missile; a fixed piece of land-based infrastructure monitored by successive Triton orbits could result in a Tomahawk cruise missile attack launched by a submarine.

'Kill Web'

And this is where the ADF seems to be starting to embrace the concept of the 'Kill Web': taking target information from intelligence sources and sensors right across the ADF and the wider Department of Defence, matching targets with the most appropriate effectors (regardless of who operates them) and then attacking the target in the most effective way the circumstances allow. A node that seems well suited to this task is the Joint Air Battle Management System (JABMS) for which Lockheed Martin has just signed the prime contract under Project AIR6500.

But JABMS isn't the only potential node in this 'Kill Web' – each of the three services and Joint Operations Command can play a significant role depending on the threat, the target and the choice of weapon.

You obviously need good voice and especially data communications to make it work; and you increasingly need Artificial Intelligence (AI) to react quickly to sudden threats or fleeting opportunities.

However, in the Australian context the AI will mostly present information to a human decision-maker because, at the end of the day, Australia's ethical approach to the use of AI and autonomous systems in combat demands that it's still a human being who makes what might be a life-threatening decision.

The targeting function and necessary communications – Link 11, Link 16, Link 21 and 22, SATCOM etc, with appropriate capacity – need to be embedded in warships, deployed land force HQs and multi-role aircraft such as the P-8A, E-7A and MC-55A.

Apart from anything else they need the autonomy to be able to engage or evade incoming missiles, for example, or to simply get on with their jobs. But the ADF's Defence Intelligence Organisation also needs to be involved for strategic and static targets such as dockyards, enemy HQs, fleets of ships and aircraft, and even satellites.

The implication of all this for the ADF is significant. In purely defensive terms, it can see further: it can spot and react to incoming weapons earlier. In an era of hypersonic missiles this matters.

But its longer reach and ability to target its weapons accurately at whatever range they can achieve also means that the ADF can make its impact felt across a much wider area. To east and west, where a potential aggressor must come across the sea or try to disrupt undersea cables, the ADF can dominate a much greater area than has been the case hitherto.

To the north and northwest of Australia lie ten separate sovereign nations – not including China – in most of which the ADF could conceivably deploy in support of a friendly government or to face a common threat. Go just a bit further afield and that number climbs to 16 nations. The ADF's combination of land-, air- and sea-launched weapons could inflict an unacceptable price on anybody attempting to misuse adjacent sea areas or a contiguous land mass.

The Australian Army's reorganisation has seen its light, Darwin-based 1st Brigade focus on littoral operations, with its much heavier Townsville-based 3rd Brigade focussed on amphibious operations – indeed one of its units, the 2nd Bn Royal Australian Regiment, is effectively Australia's amphibious battlegroup.

The clear implication is that either Brigade (depending on the terrain) or possibly both could deploy into Australia's northern archipelago and support allied and friendly governments tackling a common threat. A long-range fires (including land-based anti-ship) capability would enable island-hopping, if such a thing were necessary, and would also extend Australia's, or an ad hoc alliance's, deterrent range much further than is currently the case.

A strategy of Deterrence through Denial is enabled by weapons with a long reach, a strong, well-equipped ADF and a demonstrated willingness to use it if necessary. These are best deployed in support of Australian statecraft, including diplomacy.

This is where the Department needs to be part of a broader national security strategy, which Australia has lacked for too long. The Australian government's civil and military capabilities need to complement each other, and they must be robust and transparent. They are none of these things – at least visibly – at present

NOTES

INTRODUCTION

1. https://operationnels.com/

1. OVERVIEW TO THE 26 SEPTEMBER 2024 SIR RICHARD WILLIAMS FOUNDATION SEMINAR

1. https://sldinfo.com/2014/11/shaping-a-21st-century-approach-to-tron-warfare/

2. THE CHALLENGE FOR DEFENCE READINESS: THE IMPACT OF POLITICS

1. https://sldinfo.com/2024/06/australia-a-complacent-nation-adrift-in-the-south-pacific/

3. BUILDING COMBAT MASS: A NAVY PERSPECTIVE

1. https://defense.info/re-thinking-strategy/2024/09/maneuver-warfare-working-manned-with-autonomous-systems-for-the-ground-maneuver-element/

4. BUILDING COMBAT MASS: AN AIR FORCE PERSPECTIVE

1. https://sldinfo.com/2017/09/building-in-integration-reshaping-training-and-encompassing-development/

5. FURTHER PERSPECTIVES OF AIR VICE MARSHAL GLEN BRAZ

1. https://sldinfo.com/2017/08/group-captain-braz-on-the-raaf-and-the-way-ahead-on-electronic-warfare-shaping-a-core-distributed-capability-for-the-integrated-force/; https://sldinfo.com/2017/04/group-captain-braz-and-the-coming-of-the-growler-to-the-australian-defence-force/

8. KEY WARFIGHTING CAPABILITIES FOR THE NEW STRATEGIC ENVIRONMENT

1. https://www.directory.gov.au/portfolios/defence/department-defence/joint-capabilities-group
2. https://www.asd.gov.au/sites/default/files/2022-05/ASD-REDSPICE-Blueprint.pdf
3. https://www.asd.gov.au/about/what-we-do/redspice
4. https://defense.info/multi-domain-dynamics/2023/10/deterrence-by-denial-impactful-projection-proportionate-response-and-multi-domain-strike/

9. DEALING WITH A RAPIDLY CHANGING OPERATIONAL ENVIRONMENT

1. https://defense.info/video-of-the-week/nordic-warden-exercise-protection-of-critical-underwater-infrastructure/

10. PROTECTING THE NATION: ITS MORE THAN THE ADF'S ROLE

1. https://www.infrastructure.gov.au/infrastructure-transport-vehicles/aviation/aviation-white-paper/aviation-white-paper-terms-reference-submissions
2. https://www.defence.gov.au/about/strategic-planning/defence-industry-development-strategy
3. https://seapower.navy.gov.au/media-room/publications/chief-navy-speeches-navy-national-enterprise-0; https://strategicanalysis.org/unpacking-the-numbers-in-defences-new-integrated-investment-plan/
4. https://assets.publishing.service.gov.uk/media/5b4c60f8ed915d0927e48963/CombatAirStrategy_Lowres.pdf
5. https://airpower.airforce.gov.au/publications/airandspacepowerconference2024-transcript-resiliencethroughnetworkingandnorthernbases
6. https://www.defence.gov.au/business-industry/national-naval-shipbuilding-enterprise; https://www.industry.gov.au/news/announcing-office-national-rail-industry-coordination
7. https://sldinfo.com/2023/04/agile-basing-and-endurability-as-a-key-deterrent-capability-a-conversation-with-the-air-commander-australia/

11. THE SPACE DIMENSION

1. https://sldinfo.com/2022/07/remembering-alain-dupas-july-2022/

12. THE KEY ROLE OF DEFENCE INDUSTRY IN AUSTRALIA

1. https://news.northropgrumman.com/news/releases/northrop-grumman-australia-opens-state-of-the-art-facility
2. https://www.rtx.com/raytheon/what-we-do/integrated-air-and-missile-defense/nasams

14. INTERVIEWS SEPTEMBER-OCTOBER 2024

1. https://carnegieendowment.org/research/2024/09/us-australia-alliance-force-posture-policy-and-planning-toward-a-more-deliberate-incrementalism?lang=en
2. https://www.minister.defence.gov.au/media-releases/2024-09-05/acquisition-joint-strike-missile-boost-australias-long-range-strike-capability
3. https://sldinfo.com/2022/11/nsm-and-jsm-a-norwegian-contribution-to-the-arsenal-of-democracy/
4. https://www.defenceconnect.com.au/geopolitics-and-policy/14529-pax-americana-only-as-strong-as-the-sum-of-its-parts
5. https://navalinstitute.com.au/publications/australian-naval-review/

APPENDIX 2: THE PERSPECTIVE OF LT GENERAL STUART SIMON: PRESENTATION TO LAND FORCES 24

1. https://www.army.gov.au/news-and-events/speeches-and-transcripts/2024-09-12/chief-army-symposium-keynote-speech-human-face-battle-and-state-army-profession

APPENDIX 3: THE IMPACT OF THE AUSTRALIAN POLITICAL SYSTEM ON NATIONAL SECURITY

1. Parliament of Australia Library Briefing Book: Australia's security relationships - Australia enjoys bilateral security relationships with both New Zealand and the U.S. under ANZUS. The treaty, while not formally revoked after the U.S.- New Zealand nuclear dispute, no longer fully exists in practice. Nevertheless, a security relationship between the U.S. and New Zealand exists as members of the Five Eyes intelligence community.
2. Robert Manne, The Monthly March 2006 Essays, *Little America: How John Howard changed Australia*, https://www.themonthly.com.au/monthly-essays-robert-manne-little-america-how-john-howard-has-changed-australia-184
3. Australian Labor Party, Election 2007, Policy Document, *Labor's Plan for Defence*, November 2007.

4. Patrick Walters, *The Making of the 2009 Defence White Paper*, Security Challenges, Vol. 5, No. 2 (Winter 2009), pp. 1-10, https://www.jstor.org/stable/26459239?seq=1
5. *Defending Australia in the Asia Pacific Century: Force 2030*, Defence White Paper 2009, Commonwealth of Australia 2009, p. 64.
6. Transcripts from the Prime Minister of Australia, Australian Government, 16 November 2011, https://pmtranscripts.pmc.gov.au/release/transcript-18272
7. Jackie Calmes, *A U.S. Marine Base for Australia Upsets China, The New York Times,* 16 November 2011, https://www.nytimes.com/2011/11/17/world/asia/obama-and-gillard-expand-us-australia-military-ties.html; Phillip Coorey, John Kerin, Simon Evans, "Japanese Subs on the way," *Australian Financial Review,* 9 September, 2014, https://www.afr.com/policy/foreign-affairs/japanese-subs-on-the-way-20140908-jeqdr; Kym Bergmann, "Making sense of the Japanese submarine option," *Australian Strategic Policy Institute, The Strategist,* 9 September 2014, https://www.aspistrategist.org.au/making-sense-of-the-japanese-submarine-option/
8. *Defending Australia and its National Interests*, Defence White Paper 2013, Commonwealth of Australia 2013, p. 11.
9. Ian McPhedran, "Abbott Government to spend $20B on Japanese submarines in major blow to SA's defence industry," *news.com.au*, 7 September 2014, https://www.news.com.au/national/south-australia/abbott-government-to-spend-20-billion-on-japanese-submarines-in-major-blow-to-sas-defence-industry/news-story/070a39ca785d2d8957212e91789fa4df
10. Daniel Hurst, "Tony Abbott upbeat on China trade mission despite diplomatic tensions," *The Guardian*, 3 March 2014, https://www.theguardian.com/world/2014/mar/03/tony-abbott-upbeat-on-china-trade-mission-despite-diplomatic-tensions
11. Kerry Brown, *'Fear and Greed': A closer look at Australia's China policy*, The Diplomat, 20 April 2015, https://thediplomat.com/2015/04/fear-and-greed-a-closer-look-at-australias-china-policy/
12. Nicole Brangwin, Parliamentary Library Research Paper, *Managing SEA 1000: Australia's Attack Class Submarines*, 26 February 2020, https://www.aph.gov.au/About_Parliament/Parliamentary_Departments/Parliamentary_Library/pubs/rp/rp1920/AttackClassSubmarines
13. Naomi Woodley, *Prime Minister Abbott praises Chinese president Xi Jinping's commitment to democracy, but tourism industry not convinced by FTA*, ABC News Online, 18 November 2014, https://www.abc.net.au/news/2014-11-18/praise-for-chinese-president/5898212
14. *Defence White Paper 2016*, Commonwealth of Australia 2016, pp 43-44
15. Nichole Brangwin, Parliamentary Library Research Paper.
16. Zachary Fillingham, "Timeline: Freeze (and Thaw?) in China-Australian Relations," *Geopolitical Monitor,* 20 February 2023, https://www.geopoliticalmonitor.com/timeline-the-downward-spiral-of-china-australia-relations/
17. *Defence Strategic Update 2020*, Commonwealth of Australia, p.37
18. *Ibid.*, p. 14.

19. FRANCE24, *Were the French blindsided by the AUKUS submarine deal?,* 21 September, 2021, https://www.france24.com/en/europe/20210921-were-the-french-blindsided-by-the-aukus-submarine-deal
20. Matt Doran and Andrew Probyn, "Scott Morrison on the defensive over French submarine deal, after brief run in with Emmanuel Macron," *ABC News Online*, 31 October 2021, https://www.abc.net.au/news/2021-10-31/morrison-on-the-defen sive-over-french-subs-deal-after-brief-t%C3%AAte/100583394
21. A reference to the sinking of Rainbow Warrior, codenamed Opération Satanique, a state terrorism bombing operation by the "action" branch of the French foreign intelligence agency, the Directorate-General for External Security (DGSE), carried out on 10 July 1985.
22. Quote from "Editor's Note, Dead in the Water," Chapter 1, *Australian Foreign Affairs.*
23. Matthew Knot, "Ignore the AUKUS hand-wringers, we need these subs for sea-bed battles: Navy chief," *Sydney Morning Herald,* 15 April 2023; https://www.smh.com.au/politics/federal/ignore-the-aukus-hand-wringers-we-need-these-subs-for-sea-bed-battles-navy-chief-20230414-p5d0ev.html.
24. Soli Middelby, Anna Powles, Joanne Wallis, "AUKUS an Australia's relations in the Pacific," *East Asia Forum,* 4 November 2021, https://eastasiaforum.org/2021/11/04/aukus-and-australias-relations-in-the-pacific/
25. Soli Middelby, Anna Powles, Joanne Wallis, "AUKUS an Australia's relations in the Pacific," *East Asia Forum,* 4 November 2021, https://eastasiaforum.org/2021/11/04/aukus-and-australias-relations-in-the-pacific/
26. UK Parliament, Defence Committee Report, *UK Defence and the Indo-Pacific – Summary Report,* 24 October 2023, https://publications.parliament.uk/pa/cm5803/cmselect/cmdfence/183/summary.html
27. Margaret Reynolds, "A subservient defence policy undermines Albanese's successful first year," *Pearls and Irritations,* 22 June 2023, https://johnmenadue.com/a-subservient-defence-policy-undermines-albaneses-successful-first-year/
28. Zacchary Fillingham, *Timeline: Freeze (and Thaw?) in China-Australian Relations,* loc. cit., https://www.geopoliticalmonitor.com/timeline-the-downward-spiral-of-china-australia-relations/
29. Department of Defence Transcripts, Minister for Defence/Deputy Prime Minister Interview with Avani Dias, ABC News, 23 June 2022, https://www.minister.defence.gov.au/transcripts/2022-06-23/interview-avani-dias-abc-news
30. *Defence Strategic Review 2023*, op. cit., p. 23.
31. *Defence Strategic Review 2023*, op. cit., p. 23.
32. *Defence Strategic Review 2023* op. cit., p.18.
33. *Ibid.*, p. 45.
34. Michael Shoebridge, "New defence industry strategy: a dangerous framework that's wilfully blind," *Strategic Analysis Australia*, 2023, https://strategicanalysis.org/new-defence-industry-strategy-a-dangerous-framework-thats-wilfully-blind/
35. *Defence Strategic Review 2023*, Commonwealth of Australia, 2023, pp. 23-24
36. Institute for Integrated Economic Research-Australia, Projects, *Australia a Complacent Nation*, October 2021, https://www.jbcs.co/iieraustralia-projects

37. Department of Defence Media Release, *Release of the Defence Strategic Review*, 24 April 2023, https://www.minister.defence.gov.au/media-releases/2023-04-24/release-defence-strategic-review
38. Andrew Green, "Peter Dutton warns against UK submarines for AUKUS, drawing fire from the government," ABC News Online, 1 March 2023, https://www.abc.net.au/news/2023-03-01/peter-dutton-aukus-submarines-government-labels-irresponsible/102040234
39. Australian Foreign Affairs, "Dead in the Water: The AUKUS Delusion," AFA20, February 2024, https://www.australianforeignaffairs.com/
40. Hugh White, "Dead in the Water," *Australian Foreign Affairs*, February 2024, https://www.australianforeignaffairs.com/essay/2024/02/dead-in-the-water
41. Matthew Knott, "Defence Force under 'stress' as chief reveals true extent of staff crisis," *The Sydney Morning Herald*, 14 February 2024, https://www.smh.com.au/politics/federal/defence-force-under-stress-as-chief-reveals-true-extent-of-staff-crisis-20240214-p5f4xp.html.
42. John Hewson, "The path of political delusion," *The Saturday Paper*, 30 March 2024, https://www.thesaturdaypaper.com.au/comment/topic/2024/03/30/the-path-political-delusion
43. National Museum of Australia, Defining Moments: ANZUS Treaty, https://www.nma.gov.au/defining-moments/resources/anzus-treaty,
44. Ingeborg van Teeseling, *History Vietnam War, Australia Explained*, https://australia-explained.com.au/history/vietnam-war
45. "Australian Involvement in the Iraq War," *Wikipedia*, https://en.wikipedia.org/wiki/Australian_involvement_in_the_Iraq_War
46. Australian War Memorial, "Deaths as a result of service with Australian units," https://www.awm.gov.au/articles/encyclopedia/war_casualties.
47. Department of Health, Australian Government, "Life in Mind, Veterans and Australian Defence Force (ADF) personnel," https://lifeinmind.org.au/suicide-prevention/priority-populations/veterans-and-australian-defence-force-personnel
48. For example, "Department of Defence Transcripts, Minister for Defence/Deputy Prime Minister Television Interview," *Today Show*, 21 February 2024, https://www.minister.defence.gov.au/transcripts/2024-02-21/television-interview-today-show
49. Institute for Integrated Economic Research-Australia, Projects, *Australia a Complacent Nation*, October 2021, https://www.jbcs.co/iieraustralia-projects
50. Lili Bayer, "Europe must get ready for looming war, Donald Tusk warns," *The Guardian*, 30 March 2024.

APPENDIX 4: ADDITIONAL ANALYTICAL ASSESSMENTS

1. Defence email to author, 22 May 2024.
2. Defence email to author 20 June 2024.

ABOUT THE AUTHOR

Dr. Robbin F. Laird is a long-time analyst of global defence issues. He has worked in the U.S. government and several think tanks, including the Center for Naval Analyses and the Institute for Defense Analyses.

He is a frequent op-ed contributor to the defence press, and he has written several books on international security issues.

He is the editor of two websites, *Second Line of Defense* and *Defense.info*.

He is a member of the Board of Contributors of *Breaking Defense* and publishes there on a regular basis.

He is a research fellow with The Sir Richard Williams Foundation.

He is also based in Paris, France, and he regularly travels throughout Europe and conducts interviews with leading policy-makers in the region.

ABOUT THE INTERVIEWEES AND THE CONTRIBUTORS

VICE ADMIRAL (RETIRED) TIM BARRETT

Vice Admiral Barrett joined the Royal Australian Navy in 1976 as a Seaman Officer and later specialised in aviation. He assumed command of the Royal Australian Navy on 1 July 2014 and remained in this position until 6 July 2018.

A dual-qualified officer, Vice Admiral Barrett served in Her Majesty's Australian (HMA) Ships Melbourne, Perth and Brisbane and HMS Orkney as a Seaman Officer and then as Flight Commander in HMA Ships Stalwart, Adelaide and Canberra. His staff appointments include Deputy Director Air Warfare Development, Director Naval Officer's Postings and Director General of Defence Force Recruiting.

Vice Admiral Barrett has served as Commanding Officer 817 Squadron, Commanding Officer HMAS Albatross, Commander Australian Navy Aviation Group, Commander Border Protection Command and as Commander Australian Fleet.

Vice Admiral Barrett was awarded a Conspicuous Service Cross in 2006 for outstanding performance as Commanding Officer HMAS

Albatross and as Chief of Staff Navy Aviation Force Element Group Headquarters. Vice Admiral Barrett was appointed as a Member of the Order of Australia in 2009 and subsequently promoted to Officer of the Order of Australia in 2014 for his leadership of Border Protection Command and the Australian Fleet.

Vice Admiral Barrett holds a Bachelor of Arts in Politics and History and a Masters of Defence Studies, both from the University of New South Wales, and has completed the Advanced Management Program at Harvard Business School.

AIR VICE-MARSHAL JOHN BLACKBURN AO (RETD)

He is a former fighter pilot, test pilot, capability and strategy planner, and finally the Deputy Chief of the Royal Australian Air Force in which he served a total of 43 years.

In 2007 he was appointed an Officer in the Military Division of the Order of Australia.

He holds a Master of Arts and a Master of Defence Studies and is a life member of the Society of Experimental Test Pilots. He is the co-founder and Chair of the Institute for Integrated Economic Research Australia, a co-founder and Executive Committee member of the Australian Security Leaders Climate Group, and an editorial advisor to the French Operational SLDS (Logistics Defence Security Support) magazine.

He was formerly a Council Member of the Australian Strategic Policy Institute, Chairman of the Kokoda Foundation, Deputy Chairman of the Sir Richard Williams Foundation, and an Adjunct Professor at Edith Cowan University in Western Australia.

He also has had over 15 years experience as a consultant in the fields of Defence and National Security.

GROUP CAPTAIN ANNE BORZYCKI (RETD)

She has served a total of 35 years in the Royal Australian Air Force. During her RAAF career, Anne was seconded to Parliament House to work as the defence adviser to the Joint Standing Committee on Foreign Affairs, Defence and Trade, was a liaison officer to Capability Development Group, lead a remuneration case to change the system of the payment of flying allowance and undertook public affairs and media management roles.

She is also a co-founder and Director of the Institute for Integrated Economic Research Australia, a Fellow of the Australian Security Leaders Climate Group, and a former Board Member of the Sir Richard Williams Foundation.

DR. ANDREW CARR

Andrew Carr is a Senior Lecturer in the Strategic and Defence Studies Centre at the Australian National University.

His research focuses on Strategy and Australian Defence Policy. He has published in outlets such as Survival, Parameters, Journal of Strategic Studies, Australian Foreign Affairs, International Theory, The Washington Quarterly, Comparative Strategy. He has a sole authored book with Melbourne University Press and has edited books with Oxford University Press and Georgetown University Press.

He is currently a member of the ANU-Defence Strategic Policy History Project, writing a history of Australian Defence White Papers from 1976-2020.

GREGOR FERGUSON

Gregor Ferguson is a long-time student of innovation. An Industrial designer by training, he has studied innovation and watched innovators closely for more than three decades. His Ph.D was bestowed in

2013 by the University of Adelaide for his doctoral thesis: Product Innovation Success in the Australian Defence Industry – an Exploratory Study. He is now a defence and innovation analyst, consultant and teacher as well as a defence innovation communicator and writer.

Ferguson spent 14 years as Editor and then Editor-at-Large of *Australian Defence Magazine* (ADM), Australia's leading defence industry journal. He also wrote for the U.S. newspapers *Defense News* and *Commercial Aviation News,* for *The Australian*, and conducts specialist writing and communications for government and private sector clients.

STEPHAN FRÜHLING

Professor Stephan Frühling teaches and researches at the Strategic and Defence Studies Centre of The Australian National University and has widely published on Australian defence policy, defence planning and strategy, nuclear weapons and NATO.

Stephan was the Fulbright Professional Fellow in Australia-US Alliance Studies at Georgetown University in Washington DC in 2017. He worked as a 'Partner across the globe' research fellow in the Research Division of the NATO Defence College in Rome in 2015 and was a member of the Australian Government's External Panel of Experts on the development of the 2016 Defence White Paper.

Previously, he was the Acting Head of the Strategic and Defence Studies Centre (2022), Associate Dean Partnership and Engagement (2021-2022), Deputy Dean (2020 to 2021), and Associate Dean Education (2016 to 2020) in the ANU College of Asia and the Pacific, the inaugural Director of Studies of the ANU Master in Military Studies program at the Australian Defence Force's Australian Command and Staff College (2011 to 2013), and Managing Editor of the Kokoda Foundation's journal *Security Challenges* (2006 to 2014).

DR. MARCUS HELLYER

Marcus is Head of Research at Strategic Analysis Australia.

Marcus has spent his career addressing thorny analytical issues and wicked problems.

Previously he was a Senior Analyst at the Australian Strategic Policy Institute where he unpacked defence budget, capability and industry issues. This included demystifying the Defence Department's budget in seven editions of The Cost of Defence.

Marcus worked for 12 years in the Defence Department, primarily in its contestability function where he held several Senior Executive Service positions. This involved conducting independent capability and cost analysis of investment proposals as well as ensuring the best advice possible was provided to the Government and senior decision makers on major capital acquisitions. He also administered Defence's capital acquisition program.

Marcus has also worked in the Australian Intelligence Community.

Marcus has a first-class Honours degree in History from the University of Sydney and a Master's and Ph.D. in History and the History of Science from the University of California, San Diego. He was Assistant Professor of History at Brandeis University in Waltham, Massachusetts. He also attended the Centre for Defence and Strategic Studies where he was awarded a Master's degree in Strategic Studies.

Marcus is Director of Strategy at C2 Robotics, an Australian company specialising in the development of autonomous systems for military applications.

Marcus is also an Expert Associate at the Australian National University's National Security College.

STEPHEN KUPER

Steve has an extensive career across government, defence industry and advocacy, having previously worked for cabinet ministers at both Federal and State levels. He is a senior member of the *Defence Connect* team.

JENNIFER PARKER

Jennifer Parker **is an** Expert Associate at the Australian National University's National Security College. She has served for more than 20 years with the Royal Australian Navy (RAN). During her time in the RAN she specialised as a Principal Warfare Officer, including undertaking deep specialist training with the Royal Navy in Anti-Submarine Warfare Officer.

Jennifer has extensive operational experience from the Middle East to the Caribbean, and most areas in between. She has undertaken a series of naval appointments nationally and internationally in areas including force design strategy, career management, international exercise planning and operations.

PIERRE TRAN

Pierre Tran is a Paris-based journalist who focuses on French defence policies. He was a Reuters correspondent for many years and is a regular contributor to *Second Line of Defense* and *Defense.info*.

ADDITIONAL BOOKS ON AUSTRALIAN DEFENCE

JOINT BY DESIGN: THE EVOLUTION OF AUSTRALIAN DEFENCE STRATEGY

In the midst of the COVID-19 crisis, the prime minister of Australia, Scott Morrison, launched a new defence and security strategy for Australia. This strategy reset puts Australia on the path of enhanced defence capabilities.

The change represents a serious shift in its policies towards China, and in reworking alliance relationships going forward. "Joint by Design" is a book about Australia, but it is about the significant shift facing the liberal democracies in meeting the challenge of dealing with the 21st century authoritarian powers.

The strategic shift from land wars to full spectrum crisis management requires liberal democracies to have forces lethal enough, survivable enough, and agile enough to support full spectrum crisis management.

The book provides an overview of the evolution of Australian defence modernization over the past seven years, and the strategic shift underway.

Published December 21, 2020

AUSTRALIA AND INDO-PACIFIC DEFENCE: ANCHORING A WAY AHEAD

In Australia and Indo-Pacific Defence: Anchoring a Way Ahead, author and editor of over thirty books, Robin Laird, brings to bear his expertise on defence and security affairs to make sense of contemporary Australian international security and defence policy.

This is his third book focused on Australian defence.

It reveals the sharp mind of a person very well connected in Australian defence policy, academic and military practitioner circles. Laird has expertly sought to engage with and understand perspectives of Australian defence and security experts, many of whom are associated with the Williams Foundation, a not-for-profit Australian organisation established to advocate for the appropriate development and use of airpower, along with the other services, in defence of Australia and its interests.

This book echoes the work of the Williams Foundation which has encompassed reforms underway affecting the application not just of airpower, but also capabilities that apply to the maritime, land, space and cyber domains. It addresses the challenges of force modernisation and transformation in the context of fluctuating great power relativities (notably with the rise of an assertive and more confrontational China) in a dynamic Indo-Pacific region, at a time of significant policy initiatives affecting Australia and its place in the world.

These initiatives notably include Australia's 2023 Defence Strategic Review (DSR) and the implementation of the Australia, United Kingdom United States (AUKUS) advanced technical sharing agreement, helping Australia to acquire nuclear propulsion submarines and other advanced military capabilities.

This is an important book by a very well connected, informed and astute observer of Australia's circumstances as they pertain to defence challenges, US alliance dynamics, and technological as well as policy and political hurdles.

From the Forward by John Blaxland
Published on June 28, 2023

AUSTRALIAN DEFENCE AND DETERRENCE: A 2023 UPDATE

This is a revised edition of the book originally published in July 2023. The original book was based on the Sir Richard Williams Foundation seminar held in March 2023 which occurred between the announcement of the new submarine program and the release of the strategic defence review at the end of April 2023.

This edition adds the work done at the second 2023 Sir Richard Williams Foundation seminar held on 27 September 2023. The first dealt with the way ahead with regard to deterrence strategy; the second dealt with the expanded role of multi-domain strike as an enabler of that strategy.

The book also includes exclusive interviews as well conducted after the first and second seminars so that original interviews with both ADF personnel and leading Australian strategists are included in the book.

The book brings together in one place a contemporary historical record of Australian thinking about the reset of their defence force and way ahead in defence.

As such, it is a unique volume.

Revised edition Published in November 2023.

AUSTRALIAN DEFENCE AND DETERRENCE: A 2024 UPDATE

In this book, Dr. Laird provides an update on the ADF or the Australian Defence Force from the perspective of changes introduced in 2023 and embodied in the 2024 Defence Investment Plan. The core of the book is built around the Sir Richard Williams Foundation

April 2024 seminar on multi-domain operations in support of an Australian maritime strategy.

This book is the second in the series on Australian Defence and Deterrence the first having been published in 2023.

Laird is a research fellow with the Sir Richard Williams Foundation and has been writing their bi-annual seminar reports since 2014. And in support of these reports has conducted interviews during his visits since 2014 with the ADF and strategists about the evolution of Australia's defence and security strategy. As such, these reports provide unique insights into the evolution of Australian defence and security policy.

The book begins with a contribution by John Blackburn and Anne Borzycki who look at the broad impact of Australian politics on the nature of the Australian defence effort and argue for the need for a comprehensive national defence strategy to deal with the new historical era facing Australia and the liberal democracies.

Rather than a world of multi-polarity or great national power competition, a key aspect of the new historical epoch we have entered is multi-polar authoritarianism. Authoritarianism is clearly globally ascendent, but these regimes or groups do not share a common ideology or action program.

Many of these authoritarian states or groups have roots deeply inside Western democracies and through various means operate within Western societies, rather than simply being an external threat. The key challenge facing Australia is whether or not the government and the country shape policies and a strategy to prevail in the new historical epoch.

How does Australia generate a credible deterrent strategy against a power that is their major trading partner?

How does Australia shape a national security and defence strategy which engages the nation and mobilizes its resources?

How does Australia do so while pursuing an energy strategy which simply does not tap the natural resources which Australia possesses in abundance?

How does Australia generate a credible ADF when the government is simply putting off investments in the force that would have to fight tonite?

How credible is the future force?

Do the elements of this force really integrate or are they really new platform stove pipes?

Does Australia have a credible alliance approach?

Is AUKUS really a centerpiece of a military renaissance?

How will Australia stand up to China while largely being a raw materials supplier to the country?

How realistic is the domestic understanding of what the threat from the new authoritarians is domestically? Information war is now a key domestic fact of life, and not simply an away game.

How capable is Australia of building the defence and security infrastructure it needs?

Can Australia defend itself as a sanctuary to the extent necessary to provide a strategic reserve for its Pacific allies in times of crisis?

How will Australia focus primarily on Indo-Pacific challenges and still remain engaged with at least some presence forces for other global regions?

How will Australia defend its maritime interests without shaping a significant merchant maritime capability?

In short, key questions need to be asked and answered and not only by Australia. Each of its democratic partners faces major challenges itself.

And collectively, we face a very challenging environment with a wide range of authoritarian actors with no interest in providing their political and security capital to a "rules-based order."

How will Australia shape its way ahead?

www.ingramcontent.com/pod-product-compliance
Lightning Source LLC
Chambersburg PA
CBHW070804280326
41934CB00012B/3044

* 9 7 9 8 9 9 1 6 8 6 0 0 6 *